Lecture Notes in Computer Science 1269

Edited by G. Goos, J. Hartmanis and J. van Leeuwen

Advisory Board: W. Brauer D. Gries J. Stoer

Springer

Berlin
Heidelberg
New York
Barcelona
Budapest
Hong Kong
London
Milan
Paris
Santa Clara
Singapore
Tokyo

José Rolim (Ed.)

Randomization and Approximation Techniques in Computer Science

International Workshop RANDOM'97
Bologna, Italy, July 11-12, 1997
Proceedings

 Springer

Series Editors

Gerhard Goos, Karlsruhe University, Germany

Juris Hartmanis, Cornell University, NY, USA

Jan van Leeuwen, Utrecht University, The Netherlands

Volume Editor

José Rolim
University of Geneva, Computer Science Department
24, rue Général Dufour, CH-12011 Geneva 4, Switzerland
E-mail: rolim@cui.unige.ch

Cataloging-in-Publication data applied for

Die Deutsche Bibliothek - CIP-Einheitsaufnahme

Randomization and approximation techniques in computer science : proceedings /
International Workshop Random '97, Bologna, Italy, July 11 - 12, 1997. José
Rolim (ed.). - Berlin ; Heidelberg ; New York ; Barcelona ; Budapest ; Hong
Kong ; London ; Milan ; Paris ; Santa Clara ; Singapore ; Tokyo : Springer,
1997
 (Lecture notes in computer science ; Vol. 1269)
 ISBN 3-540-63248-4

CR Subject Classification (1991): F.2, G.1.2, G.1.6, G.2, G.3

ISSN 0302-9743
ISBN 3-540-63248-4 Springer-Verlag Berlin Heidelberg New York

Typesetting: Camera-ready by author
SPIN 10549925   06/3142 – 5 4 3 2 1 0   Printed on acid-free paper

Foreword

The Workshop on *Randomization and Approximation Techniques in Computer Science*, **Random '97**, focuses on algorithmic and complexity aspects arising in the development of efficient randomized solutions to computationally difficult problems. It aims, in particular, at fostering the cooperation among practitioners and theoreticians and among algorithmic and complexity researchers in the field. RANDOM'97, held at the University of Bologna, Italy, on July 11–12, 1997, co-locates with ICALP'97. We would like to thank the organizers of ICALP, Roberto Gorrieri, Pierpaolo Degano, and Alberto Marchetti-Spaccamela for this opportunity.

The scientific program of Random '97 consists of four sessions, namely, Approximation, Randomness, Algorithms, and Complexity. The number of papers submitted was 37. Only 14 were selected. This volume contains the selected papers plus papers by invited speakers. All papers published in the workshop proceedings were selected by the program committee on the basis of referee reports. Each paper was reviewed by at least three referees who judged the papers for originality, quality, and consistency with the themes of the conference.

We would like to thank all of the authors who responded to the call for papers, our invited speakers: S. Arora, P. Crescenzi, R. Impagliazzo, and M. Karpinski, the members of the program committee:

A. Andreev, Moscow U.
G. Ausiello, U. Roma "La Sapienza"
A. Ferreira, LIP Lyon
J. Hromkovic, U. Kiel
V. Kann, KTH Stockholm
L. Kucera, U. Prague
M. Luby, DEC/SRC
C. Papadimitriou, UC Berkeley
A. Razborov, SMI Moscow
J. Rolim, U. Geneva, chair
M. Saks, Rutgers U.
M. Serna, U. Barcelona
A. Wigderson, Hebrew U.
D. Zuckerman, UT Austin

and the reviewers Farid Ablayev, Vassil Aleksandrov, Eric Allender, Gunnar Andersson, José Balcázar, Van-Dat Cung, Lars Engebretsen, Henrik Eriksson, Bernd Gaertner, Ricard Gavaldà, Johan Hastad, Claire Kenyon, Klaus Kriegel, Danny Krizanc, Karina Marcus, Jordi Petit, Alex Russell, Alexander Shen, Luca Trevisan, Oleg Verbitsky, Peter Winkler, Laura Wynter, and Fatos Xhafa.

We gratefully acknowledge support from the European Association *INTAS* which started the *RANDOM* project and made this workshop possible. Finally, we would like to thank Danuta Sosnowska and Andrea Clementi for their help in the preparation of the meeting.

May 1997   José D. P. Rolim

Contents

Polynomial Time Approximation Schemes for Some Dense Instances of NP-Hard Optimization Problems

MAREK KARPINSKI *

Abstract. We overview recent results on the existence of polynomial time approximation schemes for some dense instances of NP-hard optimization problems. We indicate further some inherent limits for existence of such schemes for some other dense instances of the optimization problems.

1 Introduction

The computational efficiency of approximating different *NP-hard optimization* problems varies a great deal. We know by now, that unless P=NP, some problems, such as CLIQUE cannot be approximated in polynomial time even within a factor $n^{1-\epsilon}$ for any $\epsilon > 0$ (cf. Håstad [H96]). Some other problems like MAX-CUT (cf. Goemans and Williamson [GW94]) or STEINER TREE (cf. Karpinski and Zelikovsky [KZ97a]), can be approximated to within some small fixed constant factor. Till recently only a very few optimization problems were known to have polynomial time approximation schemes (PTAS), approximating to within arbitrary small constant factors.

Some of the approximation algorithms with small approximation ratios achieve also good practical performances, like some cases of STEINER TREE problems (cf. [KZ97a]), some other algorithms do not yield yet efficient practical methods of dealing with optimization problems.

In this paper we are concerned with the problem of efficient approximability of the *dense instances* of *NP-hard* optimization problems.

Recently, the first polynomial time approximation schemes have been designed *for these problems* in Arora, Karger and Karpinski [AKK95], Fernandez

* Dept. of Computer Science, University of Bonn, 53117 Bonn, Email: marek@cs.uni-bonn.de. Research partially supported by the International Computer Science Institute, Berkeley, California, by the DFG Grant KA 673 4-1, and by the ESPRIT BR Grants 7097 and EC-US 030, by DIMACS, and by the Max-Planck Research Prize.

de la Vega [FV96], Arora, Frieze and Kaplan [AFK96], and Karpinski and Zelikovsky [KZ97b]. Later on, Goldreich, Goldwasser and Ron [GGR96], and Frieze
and Kannan [FK97] gave a constant sample size approximation schemes for some
dense optimization problems. Fernandez de la Vega and Karpinski [FK97] gave
also the first polynomial time approximability characterization for dense *weighted*
instances of NP-hard problems.

This development was in contrast to the fact that the existence of such
schemes for general instances would imply that P=NP by results of Arora, Lund,
Motwani, Sudan, and Szegedy [ALMSS92].

The development above was followed by the study of the dense covering problems, Karpinski and Zelikovsky [KZ97b], and the dense bandwidth minimization
problems, Karpinski, Wirtgen and Zelikovsky [KWZ97].

It is also a very interesting artifact that the recent successes in design of the
polynomial time approximation schemes for dense optimization problems parallel the successes of the past attacks on dense approximate counting problems,
Broder [B86], Jerrum and Sinclair [JS89], Dyer, Frieze, Jerrum [DFJ94], and
Alon, Frieze and Welsh [AFW95].

2 MAX-SNP and Dense MAX-SNP Classes, and BEYOND

We consider in this Section the *dense* instances of the MAX-SNP class of optimization problems introduced by Papadimitriou and Yannakakis [PY91]. MAX-
SNP class contains constraint-satisfaction problems, where the *constraints* are
definable by quantifier-free propositional formulas.

We recall:

Definition 1. A (*maximization*) *problem A* is in MAX-SNP if there exists a sequence of relation symbols $G_1, \ldots, G_m$, a relation symbol S, and a quantifier-free
formula $\Phi(G_1, \ldots, G_m, S, x_1, \ldots, x_k)$, with x_i variables, such that the following
is true:

1. there is a polynomial time algorithm that for any given instance I of the
   problem A produces a set $\mathcal{V}$ and a sequence of relations $G_1^{\mathcal{V}}, \ldots, G_m^{\mathcal{V}}$ over $\mathcal{V}$
   ($G_i^{\mathcal{V}}$ preserve the *arity* of G_i);
2. The value of the optimum solution $OPT(I)$ of A on instance I, satisfies

$$OPT(I) =$$

$$\underset{S}{\mathrm{MAX}} \left\{ \left| \{(x_1, \ldots, x_k) \in \mathcal{V}^k \,\middle|\, \Phi(G_1^\mathcal{V}, \ldots, G_m^\mathcal{V}, S^\mathcal{V}, x_1, \ldots, x_k) = TRUE\} \right| \right\}$$

for $S^\mathcal{V}$ the relation over $\mathcal{V}$ of the same arity as S.

Example 1. MAX-CUT (cf. [GJ79], [P94]) is in MAX-SNP, since its optimum solution can be written as

$$\underset{S \subseteq \mathcal{V}}{\mathrm{MAX}} \left\{ \,|\, \{(x, y) \mid (G(x, y) \vee G(y, x) \wedge S(x) \wedge \neg S(y))\} \,|\, \right\}$$

for V the set of vertices of the graph, $G(x, y)$ its adjacency relation, and S a unary relation describing the *one side* of the cut.

For the notions of MAX-SNP-*completeness*, and MAX-SNP-*hardness* see [P94], and [AL97].

We define next the problem MAX-k-FUNCTION-SAT for some fixed integer k. MAX-k-FUNCTION-SAT has as an input m boolean functions $f_1, f_2, \ldots, f_m$ in n variables, and each f_i depends only on k variables. The problem is to find an assignment to the variables as to satisfy as many f_i's as possible.

It is known that every problem A from MAX-SNP can be viewed as a MAX-k-FUNCTION-SAT problem for a fixed k (cf. [P94]). Following [AKK95] we call an instance of a MAX-SNP problem *dense* if the corresponding instance of MAX-k-FUNCTION-SAT has $\Omega(n^k)$ functions.

Given an *optimization problem* A, a (meta) algorithm $\mathcal{A}$ is called a *polynomial time approximation scheme* (PTAS) if for every fixed $\epsilon > 0$, $\mathcal{A}$ is a *polynomial time algorithm* with *approximation ratio* $1 + \epsilon$ (meaning $\mathcal{A}$ outputs a solution S to every instance I of A such that $\mathrm{MAX}\left\{\frac{S}{OPT(I)}, \frac{OPT(I)}{S}\right\} \leq 1 + \epsilon$, for OPT(I) the optimal solution, and the running time of $\mathcal{A}$ is polynomial in the size of I).

Only a very few problems, such as KNAPSACK [IK75], and BIN PACKING [FL81], [KK82], were till recently known to have PTASs.

In [AKK95] the following general result on the existence of PTASs was proven.

Proposition [AKK95] *Dense MAX-SNP problems have PTASs.*

The proof method involves the representation of MAX-k-FUNCTION-SAT by *smooth degree-k integer* programs, and the general result on approximating such programs (cf. [AKK95]).

4

Below is the list of problems were the smooth integer programs can be applied directly to obtain the PTASs. (We call a graph *dense* if it has $\Theta(n^2)$ edges, a hypergraph of dimension d is dense if it does has $\Theta(n^k)$ edges.)

- MAX-CUT: For a given graph partition its vertices into two sets so as to maximize the number of edges between them.
- MAX-DCUT: The directed version of the MAX-CUT.
- MAX-HYPERCUT(d): A generalization of MAX-CUT to hypergraphs of dimension d (an edge is considered in a cut if it has at least one vertex on each side).
- DENSE-k-SUBGRAPH: Given a graph, find a subset of k vertices that induces a graph with the most edges (cf. [KP93]).

Following [AKK95], we have

Proposition [AKK95] *Dense instances of the following problems have PTASs: MAX-CUT, MAX-DCUT, MAX-HYPERCUT(d), and DENSE-k-SUBGRAPH for $k = \Omega(n)$.*

In what follows we call a graph G *everywhere dense* if its minimum degree is $\Omega(n)$. We consider *everywhere dense* instances of three further problems.

- SEPARATOR: Given a graph, partition its vertices into two sets, each with at least $\frac{1}{3}$ of the vertices, so as to minimize the number of edges between them.
- BISECTION: Given a graph, partition its vertices into two equal halves so as to minimize the number of edges between them.
- MIN-k-CUT: Given a graph with n vertices, and k *source* vertices, partition its vertices into k groups such that (1) each group contains one source, and (2) the number of edges between different groups is *minimized*.

Consider a graph with a *minimum* degree δn, and let c denote the capacity of its minimum bisection. The PTAS for BISECTION of [AKK95] consists of two algorithms, one of which is a PTAS when $c \geq \alpha n^2$, and the other when $c < \alpha n^2$ for α a small constant. The algorithm for $c \geq \alpha n^2$ uses the above mentioned method for approximating smooth integer programs. For the case $c < \alpha n^2$ we use the fact that in a *minimum bisection*, there must be one side whose every vertex has at most half of its neighbors on the other side, and construct a randomized exhaustive *correction* sample algorithm. The algorithm can be also easily derandomized (cf. [AKK95]). Similar PTASs work for the SEPARATOR and BISECTION problems.

Proposition [AKK95] *Everywhere dense instances of the following problems have PTASs: BISECTION, SEPARATOR and MIN-k-SAT.*

Fernandez de la Vega [FV96] has independently developed a PTAS for everywhere dense instances of MAX-CUT problem. His algorithm does not appear to generalize though to the other problems listed above.

Arora, Frieze, and Kaplan [AFK96] constructed a new rounding procedure for the quadratic assignment problem and used it to obtain PTASs on the dense instances of the NP-hard problems like QUADRATIC-ASSIGNMENT, MIN-LINEAR-ARRANGEMENT, d-DIMENSIONAL-ARRANGEMENT, BE-TWEENESS, and MIN-CUT-LINEAR-ARRANGEMENT (cf. [AFK96]).

In the other development the Regularity Lemma of Szemerédi was used to obtain more efficient PTAS for the above problems (cf. Frieze and Kannan [FK96]). Using also independent methods Goldreich, Goldwasser and Ron [GGR96], and Frieze and Kannan [FK97] gave constant time approximation schemes for some dense problems in the oracle model of computation.

3 Dense Covering Problems

We turn now to the three dense covering problems: SET COVER, STEINER TREE, and VERTEX COVER (cf. [H97], [AL97]). They do not fall into the dense MAX-SNP class definition of section 2 (VERTEX COVER is in MAX-SNP only if the degree of the graph is bounded.)

- SET COVER: Given a finite set X and a family of its subsets $\mathcal{P}$, find a minimum size subfamily $\mathcal{M}$ of $\mathcal{P}$ such that $X \subseteq \bigcup \mathcal{M}$.

  We call an instance of SET COVER ($X = \{x_1, \ldots x_n\}, \mathcal{P} = \{p_1, \ldots, p_m\}$) ϵ-*dense* (for $\epsilon > 0$) if every element of X belongs to at least ϵm sets from $\mathcal{P}$. (The instances of SET COVER are called *dense* if they are ϵ-dense for some $\epsilon > 0$. We call SET COVER restricted to dense instances a dense SET COVER accordingly.)

- STEINER TREE: Given a connected graph G and a set of its distinguished (terminal) vertices S. Find a minimum size tree within G that spans all distinguished vertices from S.

We call an instance $G = (V, E)$ of the STEINER TREE problem ϵ-*dense* if every distinguished terminal vertex is adjacent to at least $\epsilon \cdot \mid V \setminus S \mid$ nonterminals.

– VERTEX COVER: Given a graph G, find a minimum size vertex set X of G which covers all edges of G (i.e. at least one endpoint of any edge belongs to X).

We start with dense SET COVER problem. The general SET COVER was proven recently to have a threshold $(1 - o(1)) \ln n$ for the polynomial time approximation (cf. Feige [F96]) which in fact is matching asymptotically the approximation ratio by the well known greedy heuristic algorithm.

It is shown in Karpinski and Zelikovsky [KZ96] that the greedy heuristic algorithm can be applied more efficiently towards ther dense SET COVER.

Proposition [KZ96] *For any constant $c > 0$ and any $\epsilon > 0$, there is a polynomial time approximation algorithm for the ϵ-dense SET COVER with the approximation ratio $c \cdot \log n$.*

Interestingly, we cannot expect on the lower bound side of the dense SET COVER, its NP-hardness, as the results of Papadimitriou and Yannakakis [PY96] imply.

Proposition [KZ97b] *Unless $NP \subseteq DTIME [n^{\log n}]$, the dense SET COVER is not NP-hard.*

We conjecture that SET COVER cannot be approximated to within a constant factor.

Conjecture 1. The dense SET COVER cannot be approximated in polynomial time to within a constant approximation ratio.

The second problem we discuss in this section is the ϵ-dense STEINER TREE problem. We note first in passing that for $\epsilon > \frac{1}{2}$, ϵ-dense STEINER TREE problem is a special case of the network STEINER TREE problem with edge lengths 1 and 2, the problem which is still MAX-SNP-hard, Bern and Plassmann [BP89]. The best known approximation ratio for the general problem is 1.644, Karpinski and Zelikovsky [KZ97a]. For the dense STEINER TREE problem the existence of a PTAS has been recently proven in Karpinski and Zelikovsky [KZ97b].

Proposition [KZ97b] *There exists a PTAS for the ϵ-dense STEINER TREE problem.*

It is not difficult to see that there is a polynomial time reduction of the ϵ-dense SET COVER to the ϵ-dense STEINER TREE problem, and vice versa. Therefore, the similar result to Proposition 5 holds also for the dense STEINER TREE problem.

Furthermore we conjecture,

Conjecture 2. The dense STEINER TREE problem cannot be computed exactly in polynomial time .

The third problem, VERTEX COVER, is one of the first NP-hard optimization problems for which the approximation algorithms were proposed ([GJ79]). The problem is known to be MAX-SNP-hard, and the well-known 2-approximation algorithm is also believed to be the best possible (cf. [H97]). In Karpinski and Zelikovsky [KZ97b] the new approximation algorithm is designed for dense VERTEX COVER problems beating the approximation ratio 2.

We call a graph $G = (V, E)$ *everywhere ϵ-dense* if its *minimum* degree is at least $\epsilon \cdot |V|$. We call G ϵ-dense if $|E| \geq \epsilon \cdot |V|^2$.

Proposition [KZ97b] *There exists a polynomial time approximation algorithm for the VERTEX COVER problem on ϵ-dense graphs with approximation ratio $\frac{2}{2-\sqrt{1-\epsilon}}$.*

For the everywhere dense instances we get

Proposition [KZ97b] *There exists a polynomial time approximation algorithm for the VERTEX COVER problem on everywhere ϵ-dense graphs with approximation ratio $\frac{2}{1+\epsilon}$.*

Proposition 7 and 8 show that the *density* do help essentially in approximating the VERTEX COVER problem. Can we expect though existence of a PTAS for the dense VERTEX COVER problem?

The answer is no, as the *everywhere ϵ-dense* (and ϵ-dense) VERTEX COVER is MAX SNP-hard. (cf. [KZ96], [CT96]). This is due to the following *densification* construction. Start with a general instance (a graph G with n vertices) of the VERTEX COVER, and densify it by joining its all vertices with all vertices of a clique of size $\frac{\epsilon}{1-\epsilon}n$. The resulting graph is everywhere ϵ-dense. An existence of α-approximation algorithm for dense instances of VERTEX COVER entails now also $\alpha(1 + \epsilon)$-approximation algorithm for the general VERTEX COVER problem which is MAX-SNP-hard.

Proposition [CT96], [KZ96] *The dense VERTEX COVER problem is MAX-SNP-hard.*

4 Dense BANDWIDTH MINIMIZATION

We discuss now the problem of approximability of dense instances of the BAND-WIDTH problem. The BANDWIDTH problem has a long and very interesting history, and a number of important technical applications (cf., e.g. [CCDG82]). It belongs also to the class of so called layout problems and is one of hardest in this class ([DSS94]). Its approximability status resembles the BISECTION problem discussed in Section 2 in what there is a general lack of approximation algorithms with essentially sublinear approximation ratios and on the other hand it lacks also any unapproximability results. The situation on dense instances of the BANDWIDTH was even more difficult than for the dense BISECTION, for which we have constructed a PTAS (see Section 2). For the dense BANDWIDTH however, even the existence of a constant ratio approximation algorithms was an open problem. The positive result on existence of a PTAS for the dense BISEC-TION illustrates also the difficulty of proving unapproximability result for the general BISECTION problem. It indicates that the standard method of reducing *balanced* (50/50) MAX-CUT to BISECTION on the complementary graph cannot work for a good reason. The balanced MAX-CUT is MAX-SNP-hard (cf. [PY91]), however by complementing a sparse graph we get a dense one on which the BISECTION is approximable.

The situation with the BANDWIDTH is, in fact, even more subtle in this respect. The standard graph operations or a slight densification seem to destroy the structure of the instance completely.

We give now an exact formulation of the problem.

- BANDWIDTH: Given a graph $G = (V, E)$, compute the numbering of its vertices such that the maximum difference between the numbers of adjacent vertices is *minimal*.

We define also the directed BANDWIDTH problem.

– DBANDWIDTH: Given a directed graph $G = (V, E)$, compute the numbering of its vertices as above such that for every vertex v its numbering is greater than any numbering of a vertex u such that $(u, v) \in E$.

The DBANDWIDTH problem corresponds to that of minimizing the *bandwidth* of an upper triangular matrix by simultaneous row and column permutations (cf. [GGJK78]).

The problem is known to be NP-hard even if restricted to binary trees (cf. [GGJK78]), or caterpillars with hairs of length at most 3 [M83]. This makes the BANDWIDTH one of the very rare combinatorial problems which are computationally 'hard' for trees. Interestingly, the problem is efficiently computable for complete trees [Sm95]. Only a very few special cases of this problem are known to have sublinear approximation ratio algorithms, among them $\log n$-approximation algorithm for the caterpillars ([HMM91]). There are no sublinear n^ϵ-approximation algorithms known for the BANDWIDTH problem even if restricted to trees.

We consider here the BANDWIDTH problem on the everywhere dense graphs. Using a randomized placing technique combined with the special perfect matching construction Karpinski, Wirtgen and Zelikovsky [KWZ97] proved.

Proposition [KWZ97] . *There exists a randomized polynomial time approximation algorithm for the BANDWIDTH problem on everywhere dense graphs with approximation ratio 3.*

Using a more constrained nature of DBANDWIDTH the similar techniques yield.

Proposition [KWZ97] . *There exists a randomized polynomial time approximation algorithm for the DBANDWIDTH problem on everywhere dense graphs with approximation ratio 2.*

It is still an open problem whether there are constant ratio approximation algorithms for 'dense' instances of the BANDWIDTH, and the DBANDWIDTH. A challenging question remains whether there exist PTASs for the dense BANDWIDTH problems, or whether some of these problem are in fact MAX-SNP-hard.

5 Summary of Dense Approximation Results

We present here a table summarizing the results of Sections 2-4 with the best known approximation results, and the best up to date nonapproximability results on dense problems.

Problem	Approx. Ratio	Approx. Hardness	Ref.
DENSE MAX-SNP	PTAS	—	[AKK95]
DENSE MAX-CUT	PTAS	—	[AKK95],[FV96]
DENSE MAX-DCUT	PTAS	—	[AKK95]
DENSE MAX-HYPERCUT(d)	PTAS	—	[AKK95]
DENSE DENSE-K-SUBGRAPH	PTAS	—	[AKK95]
EVERYWHERE DENSE SEPERATOR	PTAS	—	[AKK95]
EVERYWHERE DENSE BISECTION	PTAS	—	[AKK95]
EVERYWHERE DENSE MIN-K-CUT	PTAS	—	[AKK95]
DENSE MIN-LINEAR-ARRANGEMENT	PTAS	—	[AFK96]
DENSE d-DIMENSIONAL-ARRANGEMENT	PTAS	—	[AFK96]
DENSE MIN-CUT-LINEAR-ARRANGEMENT	PTAS	—	[AFK96]
DENSE SET COVER	$\bigcap_c c \cdot \ln n$	OPEN	[KZ97b]
DENSE STEINER TREE	PTAS	—	[KZ97b]
DENSE VERTEX COVER	$\frac{2}{2-\sqrt{1-\epsilon}}$	MAX-SNP-hard	[KZ97b]
EVERYWHERE DENSE VERTEX COVER	$\frac{2}{1+\epsilon}$	MAX-SNP-hard	[KZ97b]
EVERYWHERE DENSE BANDWIDTH	3	OPEN	[KWZ97]
EVERYWHERE DENSE DBANDWIDTH	2	OPEN	[KWZ97]

Table 1: Table of known dense approximability results.

6 Polynomial Time Approximability of Dense Weighted Instances of NP-Hard Problems

The natural instances of optimization problems involve also weights (cf. [GJ79]) while the results studied before were concerned mainly with $0, 1$ cases. In Arora, Karger, Karpinski [AKK95], the dense MAX-CUT PTAS can be adjusted as to work also for the dense MAX WEIGHT CUT problem ([GJ79]) for the case of weights being bounded by B. In this case the algorithm produces a cut of weight at least maximum weight of a cut minus $\epsilon n^2 B$. This and also other bounded weight problems were considered briefly in Goldreich, Goldwasser and Ron [GGR96], and Frieze and Kannan [FK97]. Both papers evaluate the additional costs of handling bounded weights instead of $0,1$ weights.

In a recent paper Fernandez de la Vega and Karpinski [FK97] gave the first polynomial time approximability characterization of dense (unbounded) weighted instances of MAX WEIGHT CUT, and MAX WEIGHT BISECTION, and some other dense weighted NP-hard optimization problems, in terms of their empirical weight distribution. The crucial point of this paper is a new unbounded weight Sampling Lemma. The reader is referred to [FK97] for details.

7 Further Research and Open Problems

It remains to be seen whether the techniques used with success in the dense instances of NP-hard optimization problems, like approximating smooth higher degree integer programs by linear programs, might be useful in approximating general problems. Perhaps some other, different from exhaustive sampling methods can be developed for the nondense instances as well. Another interesting issue is to develop new more efficient techniques for the dense unbounded weight instances of the optimization problems for which costs of allowing weights are not prohibitively high.

On the level of specific dense problems discussed before, it would be interesting to shed some light on the Conjectures 1 and 2. Is there even more dramatic improvement in approximation ratio for the dense SET COVER, like $o(\log n)$, still possible (cf. Proposition 4)?

One of the most challenging open *dense* problems today is the *dense (everywhere dense)* BANDWIDTH problem. Is there an approximation ratio below 3 (cf. Proposition 10), and more strongly, is there a PTAS possible for this problem, or on the lower bound side, is this problem 'approximation hard' in some sense?

Acknowledgment

My thanks to Sanjeev Arora, Dorit Hochbaum, Haim Kaplan, Seffi Naor, and Uri Zwick for helpful discussions, and to Jürgen Wirtgen for a careful reading of the manuscript.

References

[AFW95] N. Alon, A. Frieze and D. Welsh, *Polynomial Time Randomized Approximation Schemes for the Tutte Polynomial of Dense Graphs*, Random Structures and Algorithms 6 (1995), pp. 459-478.

[AFK96] S. Arora, A. Frieze and H. Kaplan, *A New Rounding Procedure for the Assignment Problem with Applications to Dense Graph Arrangements*, Proc. 37th IEEE FOCS (1996), pp. 21-30.

[AKK95] S. Arora, D. Karger, and M. Karpinski, *Polynomial Time Approximation Schemes for Dense Instances of NP-Hard Problems*, Proc. 27th ACM STOC (1995), pp. 284-293.

[AL97] S. Arora and C. Lund, *Hardness of Approximations*, in *Approximation Algorithms for NP-Hard Problems* (D. Hochbaum, ed.), PWS Publ. Co. (1997), pp. 399-446.

[ALMSS92] S. Arora, C. Lund, R. Motwani, M. Sudan and M. Szegedy, *Proof Verification and Hardness of Approximation Problems*, Proc. 33rd IEEE FOCS (1992), pp. 14-20.

[BP89] M. Bern and P. Plassmann, *The Steiner Problem with Edge Lengths 1 and 2*, Inform. Process. Lett. 32 (1989), pp. 171-176.

[B86] A.Z. Broder, *How Hard is it to Marry at Random (On the Approximation of the Permanent)*, Proc. 18th ACM STOC (1986), pp. 50-58, Erratum in Proc. 20th ACM STOC (1988), p. 551.

[CCDG82] P. Chinn, J. Chvatalova, A. Dewdney, N. Gibbs, *The Bandwidth Problem for Graphs and Matrices - A Survey*, Journal of Graph Theory 6 (1982), pp. 223-254.

[CT96] A. Clementi and L. Trevisan, *Improved Nonapproximability Result for Vertex Cover with Density Constraints*, Proc. 2nd Int. Conference, COCOON '96, Springer-Verlag (1996), pp. 333-342.

[DSS94] J. Diaz, M. Serna and P. Spirakis, *Some Remarks on the Approximability of Graph Layout Problems*, Technical Report LSI-94-16-R, Univ. Politec, Catalunya (1994).

[DFJ94] M.E. Dyer, A. Frieze and M.R. Jerrum, *Approximately Counting Hamilton Cycles in Dense Graphs*, Proc. 4th ACM-SIAM SODA (1994), pp. 336-343.

[F96] U. Feige, *A Threshold of* ln *n for Approximating Set Cover*, Proc. 28th ACM STOC (1996), pp. 314-318.

[FV96] W. Fernandez-de-la-Vega, *MAX-CUT has a Randomized Approximation Scheme in Dense Graphs*, Random Structures and Algorithms 8 (1996), pp. 187-999.

[FK97] W. Fernandez-de-la-Vega and M. Karpinski, *Polynomial Time Approximability of Dense Weighted Instances of MAX-CUT*, Research Report No. 85171-CS, University of Bonn (1997).

[FL81] W. Fernandez-de-la-Vega, G. S. Lueker, *Bin Packing Can be Solved Within* $1+\epsilon$ *in Linear Time*, Combinatorica 1 (1981), pp. 349-355.

[FK96] A. Frieze and R. Kannan, *The Regularity Lemma and Approximation Schemes for Dense Problems*, Proc. 37th IEEE FOCS (1996), pp. 12-20.

[FK97] A. Frieze and R. Kannan, *Quick Approximation to Matrices and Applications*, Manuscript (1997).

[GGJK78] M. Garey, R. Graham, D. Johnson, D. Knuth, *Complexity Results for Bandwidth Minimization*, SIAM J. Appl. Math. 34 (1978), pp. 477-495.

[GJ79] M. R. Garey and D. S. Johnson, *Computers and Intractability: A Guide to the Theory of NP-Completeness*, W. H. Freeman (1979).

[GW94] M. Goemans and D. Williamson, *.878-approximation Algorithms for MAX-CUT and MAX2SAT*, Proc. 26th ACM STOC (1994), pp. 422-431.

[GGR96] O. Goldreich, S. Goldwasser and D. Ron, *Property Testing and its Connection to Learning and Approximation*, Proc. 37th IEEE FOCS (1996), pp. 339-348.

[HMM91] J. Haralambides, F. Makedon, B. Monien, *Bandwidth Minimization: An Approximation Algorithm for Caterpillars*, Math. Systems Theory 24 (1991), pp. 169-177.

[H96] J. Håstad, *Clique is Hard to Approximate within* $n^{1-\epsilon}$, Proc. 37th IEEE FOCS (1986), pp. 627-636.

[H97] D. Hochbaum, *Approximating Covering and Packing Problems: Set Cover, Vertex Cover, Independent Set, and Related Problems*, in *Approximation Algorithms for NP-hard Problems* (D. Hochbaum, ed.), PWS Publ. Co. (1997), pp. 94-143.

[IK75] O. H. Ibarra, C. E. Kim, *Fast Approximation Algorithms for the Knapsack and Sum of Subsets Problems*, J. ACM 22 (1975), pp. 463-468.

[JS89] M. R. Jerrum and A. Sinclair, *Approximating the Permanent*, SIAM J. Comput. 18 (1989), pp. 1149-1178.

[J94] D. S. Johnson, *Approximation Algorithms for Combinatorial Problems*, J. Comput. System Sciences 9 (1974), pp. 256-278.

[KK82] N. Karmarkar and R. M. Karp, *An Efficient Approximation Scheme for the One-dimensional Bin-Packing Problem*, Proc. 23rd IEEE FOCS (1982), pp. 312-320.

[K72] R.M. Karp, *Reducibility among Combinatorial Problems*, in Complexity of Computer Computations (R. Miller and J. Thatcher, ed.), Plenum Press (1972), pp. 85-103.

[KWZ97] M. Karpinski, J. Wirtgen and A. Zelikovsky, *An Approximation Algorithm for the Bandwidth Problem on Dense Graphs*, ECCC Technical Report TR 97-017 (1997).

[KZ96] M. Karpinski and A. Zelikovsky, *Approximating Dense Cases of Covering Problems (Preliminary Version)*, Technical Report TR-96-059, International Computer Science Institute, Berkeley (1996).

[KZ97a] M. Karpinski and A. Zelikovsky, *New Approximation Algorithms for the Steiner Tree Problem*, J. of Combinatorial Optimization 1 (1997), pp. 47-65.

[KZ97b] M. Karpinski and A. Zelikovsky, *Approximating Dense Cases of Covering Problems*, ECCC Technical Report TR 97-004, 1997, to appear in Proc. DIMACS Workshop on Network Design: Connectivity and Facilities Location, Princeton (1997).

[KP93] G. Kortsarz and D. Peleg, *On Choosing a Dense Subgraph*, Proc. 34th IEEE FOCS (1993), pp. 692-701.

[M83] B. Monien, *The Bandwidth Minimization Problem for Caterpillars with Hair Length 3 is NP-Complete*, SIAM J. Alg. Disc. Math. 7 (1986), pp. 505-514.

[P94] C. Papadimitriou, *Computational Complexity*, Addison-Wesley, (1994).

[PY91] C. Papadimitriou and M. Yannakakis, *Optmization, Approximation and Complexity Classes*, J. Comput. System Sciences 43 (1991), pp. 425-440.

[PY96] C. Papadimitriou and M. Yannakakis, *On Limited Nondeterminism and the Complexity of the VC-dimension*, J. Comput. System Sciences 53 (1996), pp. 161-170.

[R88] P. Raghavan, *Probabilistic Construction of Deterministic Algorithms: Approximate Packing Integer Programs*, J. Comput. System Sciences 37 (1988), pp. 130-143.

[Sm95] L. Smithline, *Bandwidth of the Complete k-ary Tree*, Discrete Mathematics 142 (1995), pp. 203-212.

[Y92] M. Yannakakis, *On the Approximation of Maximum Satisfiability*, Proc. 3rd ACM-SIAM SODA (1992), pp. 1-9.

Average-Case Complexity of Shortest-Paths Problems in the Vertex-Potential Model

Colin Cooper[1], Alan Frieze[2,*], Kurt Mehlhorn[3,**], and Volker Priebe[3,**]

[1] School of Mathematical Sciences, University of North London,
London N7 8DB, U.K.
[2] Department of Mathematical Sciences, Carnegie Mellon University,
Pittsburgh PA 15213, U.S.A.
[3] Max-Planck-Institut für Informatik, Im Stadtwald, D-66123 Saarbrücken, Germany.

Abstract. We study the average-case complexity of shortest-paths problems in the vertex-potential model. The vertex-potential model is a family of probability distributions on complete directed graphs with *arbitrary* real edge lengths but without negative cycles. We show that on a graph with n vertices and with respect to this model, the single-source shortest-paths problem can be solved in $O(n^2)$ expected time, and the all-pairs shortest-paths problem can be solved in $O(n^2 \log n)$ expected time.

1 Introduction

A large variety of combinatorial-optimization problems can be modeled as *shortest-paths problems*. Given a (directed) graph in which edges are assigned real edge lengths, these problems ask for *distances* between pairs of vertices. The distance of vertex j from vertex i is defined as the infimum of the lengths of all directed paths from i to j, where the length of a path is the sum of the lengths of its edges. (We need to take the infimum, since in the presence of a *negative cycle*, that is, a (directed) cycle of negative length, a finite minimum does not always exist.) We concentrate on two types of shortest-paths problems. In the *single-source shortest-paths problem*, we are interested in the distances of all vertices from a given source vertex; in the *all-pairs shortest-paths problem*, we want to compute the distance between each pair of vertices.

The *worst-case complexity* of known algorithms for the single-source shortest-paths problem on a directed graph with vertex set $[n] = \{1, \ldots, n\}$ and m edges depends heavily on whether or not edge lengths are allowed to be negative. In fact, if all edge lengths are non-negative, then Dijkstra's algorithm solves the single-source shortest-paths problem in near-linear time $O(m + n \log n)$ [7,10] and the algorithms of McGeoch [19] and Karger, Koller, and Phillips [17] solve the all-pairs shortest-paths problem in time $O(n|H| + n^2 \log n)$, where H is the set of edges that are a shortest path between their endpoints. In the general case, the Bellman–Ford algorithm [2,9] solves the single-source shortest-paths

* Research supported by NSF grant CCR-9225008.
** Research partially supported by ESPRIT LTR Project No. 20244 (ALCOM-IT).

problem in time $O(\nu m)$, where ν is the maximal number of edges on a shortest path. This is $O(nm)$ in the worst case. The solution of *one* single-source shortest-paths problem allows to transform a problem with arbitrary real edge lengths into an equivalent problem with non-negative edge lengths [8,16]. This gives a running time of $O(nm + n^2 \log n)$ for the all-pairs shortest-paths problem in the general case. Somewhat better running times are known if the edge lengths are assumed to be integers from some fixed range; see [1,12,5].

Worst-case analysis, however, sometimes fails to bring out the advantages of algorithms that perform well in practice; average-case analysis has turned out to be more appropriate for these purposes. In *average-case analysis*, we study the expected running time of shortest-paths algorithms where instances of shortest-paths problems are generated according to a probability distribution on the set of complete directed graphs with edge lengths. Two kinds of probability distributions have been considered in the literature. In the *uniform model*, the edge lengths are independent, identically distributed random variables. The *endpoint-independent model* is more general. A random instance generated according to the endpoint-independent model has the property that if the edges leaving a vertex are sorted according to their lengths, then the associated endpoints occur in random order. This even includes the case that the edge lengths themselves are arbitrarily fixed, and only the assignment of the edge lengths to the edges leaving a vertex is random. The average running time of shortest-paths algorithms has mainly been studied for the case of non-negative edge lengths. In the endpoint-independent model, the following results on the average-case complexity of the single-source shortest-paths problem (on instances with n vertices) are known. Noshita [22] analyzed the average-case complexity of Dijkstra's algorithm; the time bound, however, does not improve over the worst-case complexity of the algorithm. Spira [23] dealt first with the average-case complexity of the all-pairs shortest-paths problem. He proved an expected time bound of $O(n^2(\log n)^2)$, which was later improved by Bloniarz [3] and Frieze and Grimmett [11]. Moffat and Takaoka [21] described an algorithm with expected running time $O(n^2 \log n)$. Recently, Mehlhorn and Priebe [20] showed that the algorithm of Moffat and Takaoka is reliable, that is, it runs in time $O(n^2 \log n)$ with high probability and not just in expectation. In the uniform model, Frieze and Grimmett [11] derived precise results on the distribution of edge lengths and distances. It is a consequence of their results that the expected running time of the algorithms of McGeoch and Karger et al. is $O(n^2 \log n)$ in the uniform model.

Concerning the average-case complexity in the case of arbitrary real edge lengths, it is a major problem to define a probability distribution that allows negative edge lengths but does not trivialize the problem. For an instance of the shortest-paths problem on a directed graph D, let $D_{\geq 0}$ and $D_{<0}$ be the subgraphs formed by the edges of non-negative and negative length, respectively. If $D_{<0}$ contains a cycle and $D_{\geq 0}$ consists of a single strongly connected component, then all distances are $-\infty$. Topological sorting allows to decide in linear time whether $D_{<0}$ contains a cycle and depth-first search allows to decide in linear time whether $D_{\geq 0}$ consists of a single strongly connected component. We conclude that shortest-paths problems become trivial if the probability distribution is such that, with high probability, $D_{<0}$ contains a cycle and $D_{\geq 0}$ consists of a single strongly connected component. This will, for example, be the case if edge lengths have a non-zero probability of being negative and instances are generated according to the uniform model.

To the best of our knowledge, a probability distribution for graphs with arbitrary real edge lengths (in order to generate instances of shortest-paths problems) was proposed in the recent paper by Kolliopoulos and Stein [18] for the first time. They studied the endpoint-independent model and gave an algorithm for the single-source shortest-paths problem with expected running time $O(n^2 \log n)$. In general, it is not clear how likely a negative cycle is in the endpoint-independent model. (Note, however, that if each vertex has at least one outgoing edge of negative length, then the directed graph must contain a negative cycle.)

We study the average-case complexity of shortest-paths problems in the *vertex-potential model*. This model was used previously by Cherkassky, Goldberg, and Radzik [4] in an experimental evaluation of shortest-paths algorithms. In this model, there is a random variable π_i for each vertex $i \in [n]$ (π_i is called the *potential* of vertex i) and a random variable $r_{i,j}$ for each edge (i, j), $i, j \in [n]$. (We will set $r_{i,i} \equiv 0$, for $i \in [n]$.) The edge lengths are defined by

$$c_{i,j} = r_{i,j} - \pi_i + \pi_j \ , \ \text{for all edges } (i, j) \ .$$

Of course, only the $c_{i,j}$'s are revealed to our algorithms and the $r_{i,j}$'s and π_i's are hidden parameters of the model. We assume that the variables π_i, $i \in [n]$, are independent, identically distributed random variables with values in $[-1, 1]$ and mean 0, and we assume that the variables $r_{i,j}$, $i, j \in [n]$, $i \neq j$, are independent, identically distributed random variables with values in $[0, 1]$. (Some additional assumptions on the common distribution function F of the $r_{i,j}$'s are needed to allow the application of the results of Frieze and Grimmett [11]. This will be made more precise in Sect. 3.) The assumption $r_{i,j} \geq 0$, $i, j \in [n]$, guarantees that instances generated according to the vertex-potential model contain no negative cycle; see Propositions 1 and 2.

We show that the single-source shortest-paths problem can be solved in $O(n^2)$ expected time and that the all-pairs shortest-paths problem can be solved in $O(n^2 \log n)$ expected time. In both cases our algorithms are reliable, that is, finish their computations within the respective time bounds with high probability. Our algorithm for the single-source shortest-paths problem works in three phases. In the first phase, the algorithm computes approximations $\widehat{\pi}_i$ of the vertex potentials π_i and uses them to define reduced edge lengths $\widehat{c}$ by $\widehat{c}_{i,j} := c_{i,j} + \widehat{\pi}_i - \widehat{\pi}_j$. The reduced edge lengths will not be non-negative, they do however allow the extraction of a small number of relevant edges that, with high probability, give the same shortest-path distances. We extract $O(n^{3/2}\sqrt{\log n})$ edges and run the Bellman–Ford algorithm on this edge set. We also argue that each shortest path consists of at most $O((\log n)^2)$ edges with high probability. Taking the two facts together, we get a solution to one single-source shortest-paths problem in expected time $O(n^2)$. We use this solution to define non-negative edge lengths and then use the algorithm of McGeoch [19] or Karger, Koller, and Phillips [17] to solve the all-pairs shortest-paths problem in $O(n^2 \log n)$ expected time.

The vertex-potential model and the endpoint-independent model are incomparable as the endpoint-independent model cannot exclude negative cycles and the vertex-potential model excludes negative cycles. Moreover, in the vertex-potential model, there is a strong correlation between the endpoint of an edge and the length of the edge. Assume that the distribution function F is such that all $r_{i,j}$'s are very close to zero and that the vertex potentials π_i are uniformly distributed. This implies that for all vertices i, it is probable that the shortest edge leaving i will go to the vertex with minimum potential.

The paper is organized as follows. In Sect. 2, we recall basic facts about shortest paths and reduced edge lengths. We provide auxiliary results needed for the analysis of our algorithms in Sect. 3. The algorithms and their analysis are presented in Sect. 4.

2 Shortest Paths and Reduced Edge Lengths

Let D_n be the complete loop-less directed graph on the set $[n] = \{1, \ldots, n\}$ of vertices. Assume that edge lengths are given by a function c from the set of edges to the reals; the length of the directed edge (i, j) will be denoted by $c_{i,j}$, for $i, j \in [n]$. (For convenience, let $c_{i,i} \equiv 0$, for $i \in [n]$.) We will write (D_n, c) for D_n with edge lengths c.

For a directed path P in D_n, let $c(P)$ be the length of P with respect to c, that is, $c(P)$ is defined as $\sum_{(i,j) \in P} c_{i,j}$. For any pair i, j of vertices, let $\delta_{i,j}(c)$ be the infimum of the lengths of all paths from i to j. The quantity $\delta_{i,j}(c)$ will be referred to as the *distance* of j from i (with respect to c).

We consider the following two shortest-path problems. For a given source vertex $s \in [n]$, the single-source shortest-paths problem asks for the distances of all vertices $i \in [n]$ from s. If s is fixed and no confusion is possible, we will denote these distances by $\delta_i(c)$ for $i \in [n]$. In the all-pairs shortest-paths problem, we want to compute the distance between each pair of vertices. The maximum of all these distances is called the *diameter* of D_n with respect to c and will be denoted by $\Delta(c)$.

As mentioned in the introduction, we want to generate instances of shortest-paths problems without negative cycles. The following proposition gives a characterization of edge lengths c for which (D_n, c) does not contain a negative cycle.

Proposition 1. *The absence of negative cycles in (D_n, c) is equivalent to the existence of vertex potentials $\pi_i \in \mathbb{R}$, $i \in [n]$, so that the* reduced edge lengths r *(of c with respect to the π_i's) are non-negative, that is,*

$$r_{i,j} := c_{i,j} + \pi_i - \pi_j \geq 0 \ , \ \text{for all edges } (i, j) \ .$$

Proposition 1 relies on the following *optimality conditions* for a solution to the single-source shortest-paths problem (with source s).

For every vertex $i \in [n]$, let $d_i(c)$ denote the length (with respect to c) of some directed path from s to i, with $d_s(c) = 0$. The quantities $d_i(c)$ are equal to the distances $\delta_i(c)$ if and only if they satisfy

$$d_i(c) + c_{i,j} \geq d_j(c) \text{ , for all edges } (i,j) \text{ .} \tag{1}$$

It is well-known that shortest-paths problems are invariant under reduction of edge lengths with respect to vertex potentials. We summarize this knowledge in the following proposition.

Proposition 2. *Suppose that we associate a vertex potential $\widehat{\pi}_i \in \mathbb{R}$ with each vertex $i \in [n]$ and that we define reduced edge lengths $\widehat{c}$ (of c with respect to the $\widehat{\pi}_i$'s) by $\widehat{c}_{i,j} := c_{i,j} + \widehat{\pi}_i - \widehat{\pi}_j$ for any edge (i,j). Then, for any directed path P from vertex k to vertex ℓ,*

$$\widehat{c}(P) = \sum_{(i,j) \in P} (c_{i,j} + \widehat{\pi}_i - \widehat{\pi}_j) = c(P) + \widehat{\pi}_k - \widehat{\pi}_\ell \text{ .} \tag{2}$$

Therefore, distances with respect to c and $\widehat{c}$ relate to each other through $\delta_{k,\ell}(\widehat{c}) = \delta_{k,\ell}(c) + \widehat{\pi}_k - \widehat{\pi}_\ell$, for all $k, \ell \in [n]$. It also follows from (2) that, for any directed cycle C, $\sum_{(i,j) \in C} \widehat{c}_{i,j} = \sum_{(i,j) \in C} c_{i,j}$; in particular, $(D_n, \widehat{c})$ contains a negative cycle if and only if there is a negative cycle in (D_n, c).

3 Properties of the Vertex-Potential Model

In the vertex-potential model that we consider in this paper, edge lengths $c_{i,j}$ are generated according to

$$c_{i,j} = r_{i,j} - \pi_i + \pi_j \text{ , for all edges } (i,j) \text{ ,}$$

with random variables $r_{i,j} \geq 0$ and π_i, $i, j \in [n]$. (We will set $r_{i,i} \equiv 0$, for $i \in [n]$.) Our precise assumptions for these variables are as follows:

(A1) The variables π_i, $i \in [n]$, are independent, identically distributed random variables with values in $[-1, 1]$ and mean 0.
(A2) The variables $r_{i,j}$, $i, j \in [n]$, $i \neq j$, are independent, identically distributed random variables with values in $[0, 1]$ and mean ρ. We will denote their common distribution function by F and we assume that $F(0) = 0$ and that $F'(0)$ exists and is strictly positive. (This implies $\rho > 0$.)

The independence assumptions are the most important (and restrictive) ones. We will not need to know the value of ρ, we can get a good estimate for it from the data; see Lemma 7. This lemma will be the only place where we use that the random variables are bounded, and the assumption of boundedness is there more for convenience than necessity. It could be replaced by bounds on the tails of the distributions. By definition of $F'(0)$ and since $F(0) = 0$,

$$F(\varepsilon) = \Pr(r_{i,j} \leq \varepsilon) = F'(0) \cdot \varepsilon + o(\varepsilon) \text{ , } \quad \text{as } \varepsilon \to 0 \text{ ,} \tag{3}$$

that is, the assumption $F'(0) > 0$ implies that the distribution of the $r_{i,j}$'s can be approximated by uniform distributions in a neighborhood of 0. This allows to reduce the proof of Lemma 3 to the case of the uniform distribution on $[0, 1]$.

For a problem of size n, we will say that an event occurs *with high probability* if it occurs with probability $\geq 1 - O(n^{-\gamma})$ for an arbitrary but fixed constant $\gamma \geq 1$. To ensure a probability of failure $O(n^{-\gamma})$, in most of our statements, we have to choose sufficiently large constants, depending on the actual value of γ. This is sometimes made explicit by a subscript γ.

3.1 The Non-Negative Case ($\pi \equiv 0$)

If the $r_{i,j}$'s are distributed as in (A2) and the π_i's are identically 0, then this gives rise to an instance (D_n, r) with non-negative edge lengths, and the analysis of shortest-path algorithms by Frieze and Grimmett in [11] can be applied. We briefly review how they argue to obtain bounds on the distances in (D_n, r). For every vertex i, they construct a spanning tree T_i rooted at i. With high probability, the length of the path in T_i from the root to any other vertex is $O((\log n)/n)$ [11, (4.6) and (4.14)]. This implies that the diameter $\Delta(r)$ is $O((\log n)/n)$ with high probability. (Davis and Prieditis [6] showed that for exponentially distributed r, the expected length of a shortest path is of exactly this order of magnitude. This is also true for r uniformly distributed; see [6,19].)

Furthermore, if (i, j) is the p-th shortest edge in the adjacency list of i where $p > B_\gamma \log n$ (for a sufficiently large constant B_γ), then $r_{i,j} > \Delta(r)$ with high probability [11, Lemma 4.3]. This means that, with high probability, edge (i, j) is not contained in any shortest path in (D_n, r). (Edges (i, j) with $r_{i,j} > \Delta(r)$ are *irrelevant* for shortest-paths computations.) We can therefore restrict ourselves to examining the sparse graph with only the $O(\log n)$ shortest edges from each adjacency list. We will show in Lemma 8 how to adjust this idea of sparsifying the graph if edge lengths are distributed according to the vertex-potential model.

Lemma 3 ([11]). *Suppose that edge lengths in (D_n, r) are distributed as specified in (A2). Then, with high probability, $\Delta(r) = O((\log n)/n)$, and the set $A := \{(i, j) \; ; \; r_{i,j} \leq \Delta(r)\}$ of relevant edges is of cardinality $O(n \log n)$.*

(For the special case of uniformly distributed r, similar results were proved by Hassin and Zemel in [14].) By a fairly involved argument, Frieze and Grimmett also proved that each of the trees T_i, $i \in [n]$, is of depth $O(\log n)$, that is, that the short (in length, though not necessarily shortest-length) paths in T_i consist of $O(\log n)$ edges with high probability [11, Theorem 5.2]. *Shortest* paths may have more (but shorter) edges, but we now prove that, with high probability, shortest paths also consist of few edges only.

Lemma 4. *With high probability, shortest paths in (D_n, r) consist of $O((\log n)^2)$ edges.*

Proof. This will follow from the fact that if at least $|P|/k$ of the edges of a shortest path P in (D_n, r) have length at least α, then $|P| \leq k \cdot \Delta(r)/\alpha$, where

$|P|$ denotes the number of edges in P. By (3), we can fix $\alpha = \Theta(1/n)$ so that $p_\alpha = \Pr(r_{i,j} \leq \alpha) \leq 1/(2n)$ for sufficiently large n; an edge (i,j) with $r_{i,j} \leq \alpha$ will be called *tiny*. For any $k \geq 1$, the probability that a (directed) path with k edges consists only of tiny edges is $\leq p_\alpha^k$. Hence, with probability $\geq 1 - 2\binom{n}{k+1}p_\alpha^k \geq 1 - n/2^k$, no path in (D_n, r) with $\geq k$ edges consists only of tiny edges, which implies that any path P in (D_n, r) has at least $\lfloor |P|/k \rfloor$ edges of lengths $> \alpha$. If we set $k = K_\gamma \log n$ with K_γ chosen large enough, then, together with the bound on $\Delta(r)$ from Lemma 3, we conclude that $|P| = O((\log n)^2)$ for any shortest path P in (D_n, r) with high probability. □

Since shortest paths are invariant under reduction of edge lengths with respect to vertex potentials (see Sect. 2), the following is an immediate consequence of Lemma 4.

Corollary 5. *Let edge lengths r be distributed as specified in (A2), and let edge lengths c be obtained by reducing the edge lengths r with respect to some vertex potentials. Shortest paths in (D_n, c) consist then of $O((\log n)^2)$ edges with high probability.*

3.2 The General Case: Approximating the Vertex Potentials

We will use the following form of the well-known *Chernoff–Hoeffding bound* on the tail of the distribution of a sum of independent random variables; see [15,13] for a proof.

Lemma 6. *Let X be the sum of independent, identically distributed random variables $X_1, \ldots, X_m$ with values in $[0,1]$; let $\xi := E[X_1]$. Then, for any ε, $0 < \varepsilon < 1$,*

$$\Pr(|X/m - \xi| > \varepsilon\xi) \leq 2 \cdot e^{-\varepsilon^2 m\xi/3} . \tag{4}$$

If the X_i's take values in $[a,b]$, then, by a simple linear transformation, we get

$$\Pr(|X/m - \xi| > \varepsilon(\xi - a)) \leq 2 \cdot e^{-\varepsilon^2 m(\xi-a)/(3(b-a))} . \tag{5}$$

We now show how to compute vertex potentials $\widehat{\pi}_i$, $i \in [n]$, that are good approximations to the random variables π_i, $i \in [n]$, with high probability.

Lemma 7. *For any $i \in [n]$, define*

$$\widehat{\pi}_i := \widehat{\rho} - \frac{1}{n-1} \cdot \sum_{j=1}^n c_{i,j} , \quad \text{where} \quad \widehat{\rho} := \frac{1}{n(n-1)} \sum_{i,j=1}^n c_{i,j} .$$

Then, with high probability, $|\widehat{\pi}_i - \pi_i|$ is of order $O(\sqrt{(\log n)/n})$.

Proof. For any $i \in [n]$, by definition of $c_{i,j}$,

$$\frac{1}{n-1} \cdot \sum_{j=1}^n c_{i,j} = \frac{1}{n-1} \cdot \sum_{j=1}^n (r_{i,j} - \pi_i + \pi_j) = -\pi_i + \frac{1}{n-1} \cdot \sum_{\substack{j=1 \\ j\neq i}}^n \pi_j + \frac{1}{n-1} \cdot \sum_{j=1}^n r_{i,j} ,$$

which implies that

$$\widehat{\pi}_i - \pi_i = \widehat{\rho} - \frac{1}{n-1} \cdot \sum_{j=1}^{n} r_{i,j} - \frac{1}{n-1} \cdot \sum_{\substack{j=1 \\ j \neq i}}^{n} \pi_j \ .$$

Hence,

$$|\widehat{\pi}_i - \pi_i| \leq |\widehat{\rho} - \rho| + \left| \frac{1}{n-1} \cdot \sum_{j=1}^{n} r_{i,j} - \rho \right| + \left| \frac{1}{n-1} \cdot \sum_{\substack{j=1 \\ j \neq i}}^{n} \pi_j \right| \ . \tag{6}$$

Recall that $E[\sum_{j \neq i} \pi_j] = 0$ and $\frac{1}{n-1} \cdot E[\sum_j r_{i,j}] = \rho$, and that we can bound the deviation from the expected values for both $\sum_{j \neq i} \pi_j$ and $\frac{1}{n-1} \cdot \sum_j r_{i,j}$ by means of the Chernoff–Hoeffding bounds (5) and (4), respectively. More precisely, we get that, with high probability,

$$\left| \tfrac{1}{n-1} \cdot \sum_{j \neq i} \pi_j \right| = O\left(\sqrt{(\log n)/n} \right) \ , \tag{7}$$

$$\left| \tfrac{1}{n-1} \cdot \sum_j r_{i,j} - \rho \right| = O\left(\sqrt{(\rho \cdot \log n)/n} \right) = O\left(\sqrt{(\log n)/n} \right) \ . \tag{8}$$

Furthermore, $\widehat{\rho}$ is a very good estimate for ρ. Since

$$\widehat{\rho} := \frac{1}{n(n-1)} \sum_{i,j=1}^{n} c_{i,j} = \frac{1}{n(n-1)} \sum_{i,j=1}^{n} r_{i,j} \ ,$$

$\widehat{\rho}$ is the sum of $n(n-1)$ independent, identically distributed random variables with values in $[0,1]$, and $E[\widehat{\rho}] = \rho$. By (4), we get that, with high probability,

$$|\widehat{\rho} - \rho| = O\left(\sqrt{(\rho \cdot \log n)/(n(n-1))} \right) = O\left(\sqrt{\log n}/n \right) \ . \tag{9}$$

By (9), (8), and (7), each term on the right-hand side in (6) is $O(\sqrt{(\log n)/n})$ with high probability, which implies that $|\widehat{\pi}_i - \pi_i|$ is $O(\sqrt{(\log n)/n})$ with high probability. $\qquad \square$

4 The Algorithms

4.1 Solving the Single-Source Shortest-Paths Problem

We are now ready to explain in detail how we solve an instance (D_n, c) of the single-source shortest-paths problem if edge lengths in (D_n, c) are generated according to the vertex-potential model of Sect. 3. Our algorithm proceeds in three phases, a preprocessing phase, a computation phase, and a postprocessing phase, in which the correctness of the solution from the second phase is checked. Let the source vertex for the single-source shortest-paths problem under consideration be denoted by s.

Preprocessing. The algorithm computes, for every vertex $i \in [n]$, a vertex potential $\widehat{\pi}_i$ as in Lemma 7 and transforms the edge lengths $c_{i,j}$ to

$$\widehat{c}_{i,j} := c_{i,j} + \widehat{\pi}_i - \widehat{\pi}_j = r_{i,j} + (\widehat{\pi}_i - \pi_i) - (\widehat{\pi}_j - \pi_j) , \text{ for all edges } (i,j) . \quad (10)$$

The shortest-paths problems (D_n, c) and $(D_n, \widehat{c})$ are equivalent. However, $(D_n, \widehat{c})$ is more efficiently solvable, since edge lengths $\widehat{c}$ allow a substantial sparsification of the underlying graph. The following lemma gives a bound on the length of irrelevant edges (with respect to $\widehat{c}$).

Lemma 8. *If $\widehat{c}_{i,j} > \Delta(r) + 2\max_i |\widehat{\pi}_i - \pi_i|$, then edge (i,j) is not contained in any shortest path.*

Proof. If (i,j) is contained in a shortest path, then $\widehat{c}_{i,j} = \delta_{i,j}(\widehat{c}) \leq \Delta(\widehat{c})$. It follows from (10) and Proposition 2 that $\delta_{k,\ell}(\widehat{c}) = \delta_{k,\ell}(r) + (\widehat{\pi}_k - \pi_k) - (\widehat{\pi}_\ell - \pi_\ell)$, for any $k, \ell \in [n]$. This implies $\Delta(\widehat{c}) \leq \Delta(r) + 2\max_i |\widehat{\pi}_i - \pi_i|$. □

For $\gamma \geq 1$ arbitrary but fixed, we know from Lemma 3 and from Lemma 7 that there exist constants C_γ and M_γ so that

$$\Delta(r) \leq C_\gamma (\log n)/n \quad \text{and} \quad 2\max_i |\widehat{\pi}_i - \pi_i| \leq M_\gamma \sqrt{(\log n)/n} \quad (11)$$

with probability $\geq 1 - O(n^{-\gamma})$. For the time being, we will assume that (11) holds. For an arbitrary (but fixed) constant $L_\gamma > M_\gamma$ (and sufficiently large n), all *relevant* edges (that is, edges that will possibly be contained in a shortest path) are then contained in

$$\widehat{A} := \left\{ (i,j) ; \ \widehat{c}_{i,j} \leq L_\gamma \sqrt{(\log n)/n} \right\} .$$

(A more careful proof of Lemma 7 reveals that we could set $L_\gamma = 12\sqrt{\gamma}$. This value is not optimal but nevertheless indicates that, for given γ, some explicit constant L_γ is easily derivable.) The vertex potentials $\widehat{\pi}_i$, $i \in [n]$, the reduced edge lengths $\widehat{c}$, and the set $\widehat{A}$ can be computed in $O(n^2)$ time. Let $\widehat{D}$ denote the graph $([n], \widehat{A})$.

Computation. We now solve a single-source shortest-paths problem with source s on the sparsified graph $(\widehat{D}, \widehat{c})$ by running the Bellman–Ford algorithm [2,9]. This algorithm maintains tentative distances d_i for every vertex $i \in [n]$. The d_i's are initially set to ∞ (except for $d_s = 0$), and d_i always represents the length of some path in $(\widehat{D}, \widehat{c})$ from s to i. The Bellman–Ford algorithm proceeds in passes over the edge set $\widehat{A}$, maintaining the following invariant. After the k-th pass, the Bellman–Ford algorithm has correctly computed the distances of all vertices to which there is a shortest path from s consisting of at most k edges. The algorithm actually checks the optimality conditions (1) for all edges (in $\widehat{A}$) in each pass, and it will therefore terminate (with all distances in $(\widehat{D}, \widehat{c})$ computed correctly) after the ν-th pass, where ν is the maximum number of edges in a shortest path in $(\widehat{D}, \widehat{c})$. The running time of the Bellman–Ford algorithm is therefore $O(\nu \cdot |\widehat{A}|)$, which is $O(n^3)$ in the worst case but $o(n^2)$ with high probability, as we now argue.

By (10) and (11),

$$\widehat{A} \subseteq \left\{ (i,j) \; ; \; r_{i,j} \leq (L_\gamma + M_\gamma)\sqrt{(\log n)/n} \right\} ,$$

and it follows from (3) that edge (i,j) is an element of the set on the right-hand side with probability $\widehat{p} = \Theta(\sqrt{(\log n)/n})$ for sufficiently large n, independently of all the other edges. The random variable $|\widehat{A}|$, the cardinality of $\widehat{A}$, is therefore stochastically dominated by a random variable that is binomially distributed with parameters $n(n-1)$ and $\widehat{p}$. We apply the tail estimates of Lemma 6 to deduce that $|\widehat{A}| = O(n(n-1)\sqrt{(\log n)/n}) = O(n^{3/2}\sqrt{\log n})$ with high probability. By (10) and Corollary 5, $\nu = O((\log n)^2)$ with high probability. Hence, with high probability, it takes $O(n^{3/2}(\log n)^{5/2}) = o(n^2)$ time to run the Bellman-Ford algorithm on the sparsified graph $(\widehat{D}, \widehat{c})$.

Postprocessing. The second phase will have failed to compute all distances correctly only if (11) does not hold, which happens only with probability $O(n^{-\gamma})$. The optimality conditions (1) (checked for all edges) are an $O(n^2)$-time certificate for the correctness of the solution. Since the worst-case running time of the Bellman-Ford algorithm on $(D_n, \widehat{c})$ is $O(n^3)$, we can easily afford to run the Bellman–Ford algorithm on $(D_n, \widehat{c})$ in case of failure, without affecting the bounds on the running time. Finally, it takes $O(n^2)$ time to compute the distances $\delta_i(c) = \delta_i(\widehat{c}) - \widehat{\pi}_s + \widehat{\pi}_i$, $i \in [n]$.

The discussion above is summarized in the following theorem.

Theorem 9. *Assume that edge lengths in (D_n, c) are generated according to the vertex-potential model of Sect. 3. The single-source shortest-path problem can then be solved in time $O(n^2)$ with high probability.*

4.2 Solving the All-Pairs Shortest-Paths Problem

Theorem 10. *Assume that the edge lengths in (D_n, c) are generated according to the vertex-potential model of Sect. 3. The all-pairs shortest-path problem can then be solved in $O(n^2 \log n)$ with high probability.*

Proof. We first compute distances $\delta_i(c)$, $i \in [n]$, with respect to source vertex 1 by solving a single-source shortest-paths problem as in the proof of Theorem 9. This takes time $O(n^2)$ with high probability. Let $\bar{c}$ be the reduced edge lengths of c with respect to the vertex potentials $\delta_i(c)$, $i \in [n]$, that is, for all edges (i,j),

$$\begin{aligned} \bar{c}_{i,j} &= c_{i,j} + \delta_i(c) - \delta_j(c) \\ &= r_{i,j} + (\delta_i(c) - \pi_i) - (\delta_j(c) - \pi_j) . \end{aligned} \qquad (12)$$

It follows from the first equality and the optimality conditions (1), that $\bar{c}_{i,j} \geq 0$ for all edges (i,j). The reduced edge lengths $\bar{c}$ can be computed in $O(n^2)$ time, and the same time bound will allow to transform distances $\delta_{i,j}(\bar{c})$ into distances $\delta_{i,j}(c)$, for all $i, j \in [n]$.

To compute the $\delta_{i,j}(\bar{c})$'s, we run one of the algorithms of Karger, Koller, and Phillips [17] or McGeoch [19] on $(D_n, \bar{c})$ that efficiently solve the all-pairs shortest-paths problem with non-negative edge lengths. Both algorithms run in time $O(n^2 \log n + n|H|)$ where $H = H(\bar{c})$ is the set of edges that are the shortest path (with respect to $\bar{c}$) between their endpoints. We apply the arguments of Sect. 2 again. Shortest paths are invariant under reduction of the edge lengths with respect to vertex potentials, and it follows from (12) that the edges in H are also the shortest path between their endpoints with respect to edge lengths r. H is therefore contained in the set $A := \{(i,j) \ ; \ r_{i,j} \leq \Delta(r)\}$, and it follows by Lemma 3 that $|H| = O(n \log n)$ with high probability. This yields a running time of $O(n^2 \log n)$ with high probability for the algorithms by McGeoch and Karger, Koller, and Phillips. Since $|H| = O(n^2)$ in the worst case, our algorithm has a running time of $O(n^3)$ with probability $O(n^{-\gamma})$, $\gamma \geq 1$, which still gives an expected running time of $O(n^2 \log n)$. □

References

1. R. K. Ahuja, K. Mehlhorn, J. B. Orlin, and R. E. Tarjan, Faster algorithms for the shortest path problem, *J. Assoc. Comput. Mach.* **37** (1990) 213–223
2. R. Bellman, On a routing problem, *Quart. Appl. Math.* **16** (1958) 87–90
3. P. A. Bloniarz, A shortest-path algorithm with expected time $O(n^2 \log n \log^* n)$, *SIAM J. Comput.* **12** (1983) 588–600
4. B. V. Cherkassky, A. V. Goldberg, and T. Radzik, Shortest paths algorithms: Theory and experimental evaluation, *Math. Programming* **73** (1996) 129–174
5. B. V. Cherkassky, A. V. Goldberg, and C. Silverstein, Buckets, heaps, lists, and monotone priority queues, *Proc. 8th Annual ACM-SIAM Symposium on Discrete Algorithms*, New Orleans LA, 1997, 83–92
6. R. Davis and A. Prieditis, The expected length of a shortest path, *Inform. Process. Lett.* **46** (1993) 135–141
7. E. W. Dijkstra, A note on two problems in connexion with graphs, *Numer. Math.* **1** (1959) 269–271
8. J. Edmonds and R. M. Karp, Theoretical improvements in algorithmic efficiency for network flow problems, *J. Assoc. Comput. Mach.* **19** (1972) 248–264
9. L. R. Ford, Jr. and D. R. Fulkerson, *Flows in Networks*, Princeton University Press, Princeton NJ, 1962
10. M. L. Fredman and R. E. Tarjan, Fibonacci heaps and their uses in improved network optimization algorithms, *J. Assoc. Comput. Mach.* **34** (1987) 596–615
11. A. M. Frieze and G. R. Grimmett, The shortest-path problem for graphs with random arc-lengths, *Discrete Appl. Math.* **10** (1985) 57–77
12. A. V. Goldberg, Scaling algorithms for the shortest paths problem, *SIAM J. Comput.* **24** (1995) 494–504
13. T. Hagerup and C. Rüb, A guided tour of Chernoff bounds, *Inform. Process. Lett.* **33** (1989/90) 305–308
14. R. Hassin and E. Zemel, On shortest paths in graphs with random weights, *Math. Oper. Res.* **10** (1985) 557–564

15. W. Hoeffding, Probability inequalities for sums of bounded random variables, *J. Amer. Statist. Assoc.* **58** (1963) 13–30

16. D. B. Johnson, Efficient algorithms for shortest paths in sparse networks, *J. Assoc. Comput. Mach.* **24** (1977) 1–13

17. D. R. Karger, D. Koller, and S. J. Phillips, Finding the hidden path: Time bounds for all-pairs shortest paths, *SIAM J. Comput.* **22** (1993) 1199–1217

18. S. G. Kolliopoulos and C. Stein, Finding real-valued single-source shortest paths in $o(n^3)$ expected time, in: W. H. Cunningham, S. T. McCormick, and M. Queyranne (Eds.), *Integer Programming and Combinatorial Optimization* (Lecture Notes in Computer Science; 1084), Springer-Verlag, Berlin, 1996, 94–104

19. C. C. McGeoch, All-pairs shortest paths and the essential subgraph, *Algorithmica* **13** (1995) 426–441

20. K. Mehlhorn and V. Priebe, On the all-pairs shortest-path algorithm of Moffat and Takaoka, *Random Structures Algorithms* **10** (1997) 205–220

21. A. Moffat and T. Takaoka, An all pairs shortest path algorithm with expected time $O(n^2 \log n)$, *SIAM J. Comput.* **16** (1987) 1023–1031

22. K. Noshita, A theorem on the expected complexity of Dijkstra's shortest path algorithm, *J. Algorithms* **6** (1985) 400–408

23. P. M. Spira, A new algorithm for finding all shortest paths in a graph of positive arcs in average time $O(n^2 \log^2 n)$, *SIAM J. Comput.* **2** (1973) 28–32

Approximation Algorithms for Covering Polygons with Squares and Similar Problems

(Extended Abstract)

Christos Levcopoulos   Joachim Gudmundsson

Department of Computer Science
Lund University, Box 118, S-221 00 Lund, Sweden

Abstract. We consider the problem of covering arbitrary polygons, without any acute interior angles, using a preferably minimum number of squares. The squares must lie entirely within the polygon.

Let P be an arbitrary input polygon, with n vertices, coverable by squares. Let $\mu(P)$ denote the minimum number of squares required to cover P. In the first part of this paper we present an algorithm which guarantees a constant (14) approximation factor running in $O(n^2 + \mu(P))$ time. As a corollary we obtain the first polynomial-time, constant-factor approximation algorithm for "fat" rectangular coverings. In the second part we show an $O(n \log n + \mu(P))$ time algorithm which produces at most $11n + \mu(P)$ squares to cover P. In the hole-free case this algorithm runs in linear time and produces a cover which is within an $O(\alpha(n))$ approximation factor of the optimal, where $\alpha(n)$ is the extremely slowly growing inverse of Ackermann's function. In parallel our algorithm runs in $O(\log n)$ randomized time using $O(\max(\mu(P), n))$ processors.

1 Introduction

One of the main topics of computational geometry is how to decompose polygonal objects into simpler polygons, such as triangles, squares, rectangles, convex polygons and star-shaped polygons [15]. We distinguish between two types of decomposition, *partitions* and *coverings*. A decomposition is called a partition if the object is decomposed into non-overlapping pieces. If the pieces are allowed to overlap, then we call the decomposition a covering.

In this paper we consider the problem of covering polygons with squares. Given a polygon P, the *square covering* problem asks for a minimum number of (possibly overlapping) squares whose union is P.

Decomposition problems have many applications both within and outside computer science [7,10]. When solving other problems for geometric figures, a common method is to decompose the figure into simpler parts, solve the problem on each component using a specialized algorithm, and then successively combine the partial solutions to solve the problem for larger and larger parts of the polygon [15].

O'Rourke and Supowit [14] showed that the problems of covering polygons with a minimum number of convex polygons, star-shaped polygons or spiral polygons are all NP-hard, if the polygon contains holes.

The related *rectilinear square covering* problem, i.e. when the polygons, as well as the squares, have sides that are vertical or horizontal, has been treated in several papers [1,2,13,17]. Some of the previous work on the rectilinear square cover problem used a bit-map representation for the input polygon [1,13,17]. A *bit-map* representation of a polygon is a boolean (zero-one) matrix, where one (1) represents a point inside the polygon and a zero (0) - a point outside it. When using this representation, complexity is measured in terms of the number of points in the matrix, denoted p. Note that $p > \Omega(n)$, and for most practical applications $p \gg n$, where n is the number of edges of P. Bar-Yehuda and Ben-Hanoch argues that the representation of the polygon should be made as segment representation, not only theoretically, but also for most practical applications [2].

Scott and Iyenger [17] present an algorithm to find, in $O(n \log n)$ time, the maximal squares in the polygon, and then divide the rectangular portions of the polygon to cover them minimally. However, their algorithm does not yield a globally minimum cover. It was shown by Aupperle, Conn, Keil and O'Rourke [1] that the rectilinear case is NP-hard for polygons containing holes. In the case where the image is hole-free, they provide an $O(p^{1.5})$ algorithm.

Recently, Bar-Yehuda and Ben-Hanoch [2] presented a linear time algorithm for covering simple (hole-free) rectilinear polygons with squares. Morita [13] developed a parallel algorithm, which finds a *minimal* (not minimum) square cover for bit-maps which may contain holes. The sequential running time of this algorithm is $O(p)$. A square cover is called *minimal* if it has no smaller subset that forms a cover. A square cover is called *minimum* if there is no smaller set that forms a cover.

The *general rectangle cover* problem has also been treated in several papers [3,5,7,11,12]. One application for this problem is the fabrication of masks for VLSI chips. In [11] an algorithm was presented which covers a polygon P with $O(n \log n + \mu'(P))$ rectangles in $O(n \log n + \mu'(P))$ time, where $\mu'(P)$ is the minimum number of rectangles needed to cover P. In [5] a different heuristic was presented, guaranteeing an $O(\log n)$ approximation factor in polynomial time $(\Omega(n^6))$, provided that the vertices of the polygon have polynomially bounded integer coordinates. Even if the rectangles have bounded aspect ratio, i.e. the ratio between the longest and the shortest side of the rectangles is bounded by a constant, then, prior to this paper, no polynomial-time approximation algorithm was known to guarantee an approximation factor better than $O(\log n)$.

Thus, in this paper we show the first constant-factor approximation algorithm for rectangles with bounded aspect ratio running in time $O(n^2 + \mu^*(P))$, where $\mu^*(P)$ denotes the minimum number of (bounded ratio) rectangles needed to cover an arbitrary polygon P. We present two algorithms which cover an arbitrary polygon P, without any acute interior angles, with squares. In section 2 we present a new square covering algorithm which guarantees a constant approximation factor running in $O(n^2 + \mu(P))$ time. We obtain the same results for covering with fat rectangles, but the approximation factors increase by some constant, depending on how fat the rectangles are (Section 4). In section 3, we present an algorithm, by refining the technique proposed in [11], which covers an arbitrary polygon P with at most $11n+\mu(P)$ squares in time $O(n \log n+\mu(P))$. If the polygon is hole-free, then the algorithm produces a covering which is within an $O(\alpha(n))$ approximation factor in optimal time $O(n+\mu(P))$. The algorithm works by partitioning the polygon into smaller pieces, which are then covered independently.

2 Achieving a Constant Approximation Factor

In this section we will present a simple algorithm, H, which covers an arbitrary polygon P with at most $24 \times \mu(P)$ squares in quadratic time. We also show the modifications that are needed to improve the approximation factor from 24 to 14 (Section 2.1). Let $H(P)$ denote the number of squares produced by H. Let $H_1, \ldots, H_5$ denote the five different steps of H, and let $H_2(P), H_5(P)$ denote the number of squares produced in steps 2 and 5.

Step 1: Generate a list of Voronoi faces (H_1)
The first step of our algorithm is to construct a list of all faces of the generalized Voronoi diagram of P. Each face is represented by a list of its edges in clockwise order. In [9] the notion of the Voronoi diagram, also called the polygonal skeleton or medial axis, is generalized to include open line segments as well as points. Given an arbitrary polygon P, the part of the generalized Voronoi diagram lying within P is constructed which partitions P into $n+w$ faces, Fig. 1a, where each segment and each concave vertex induces a face of the Voronoi diagram (w is the number of concave vertices of P). Every point lying in a face induced by d, where d is either a segment or a concave vertex, lies at least as close to d as to any other point of the boundary of P. By using Kirkpatrick's algorithm the time complexity of H_1 is $O(n \log n)$.

Step 2: Detect and cover all funnels (H_2)
In this step we need some previous results and definitions.

Definition 1. [11] Let P be any hole-free polygon. A *funnel cell* of P is a trapezoidal piece of a Voronoi face in P having the following properties. Let A, B, C and D be the vertices of the trapezoid in counter-clockwise order, such that AD is parallel to and shorter than BC, and the interior angle (A, D, C) is greater than or equal to 90 and less than 135 degrees, Fig. 1b. The segment AB lies on an edge of P, say e, and the segment CD is a Voronoi edge bounding the Voronoi face induced by e in P [9].

By this, and by the definition of generalized Voronoi diagrams, it follows that the mirror image of a funnel cell with respect to the Voronoi edge bounding the cell is also a funnel cell. Such a pair of funnel cells with a Voronoi edge separating them is called a *funnel*, Fig. 1b. Let A' and B' be the symmetric image of A respectively B.

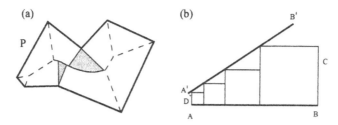

Fig. 1. (a) Medial axis (dashed). Faces induced by concave vertices are shadowed. (b) A funnel.

Observation 2. *Let F denote the set of all funnels in an arbitrary polygon P.*
a) It holds that $\mu(P)=O(n+\sum_{t\in F}\mu(t))$, where $\mu(t)$ is the minimum number of squares needed to cover a funnel t.
b) All funnels of P, if there are any, can be detected in $O(n\log n)$ time [12].

Since we can find all the funnels in P in $O(n\log n)$ time, we will start the covering procedure (Steps 2 and 5) by covering all funnels. Let t be any funnel of F, and let A, B, A' and B' be the vertices of the funnel in counter-clockwise order, Fig. 1b. We may assume w.l.o.g. that AB is horizontal and at the bottom and that $|BB'|\geq|AA'|$. We start by placing a square s_1 on AB with lower left corner at A and upper left corner at $A'B'$, then place a square s_2 on AB with lower left corner at s_1's lower right corner and upper left corner on $A'B'$. Continue as described until AB is entirely included by the produced set of squares. Let S denote the set of squares, $s_1,\ldots,s_m$, produced to include AB. If $A'B'$ is also included in S, i.e. if the interior angle $\angle ADC$ is $90°$, then we are finished, otherwise we do the corresponding procedure on $A'B'$.

Since the squares in a covering of P have to include all the edges of P, it is easily seen that the number of squares produced in this step is optimal. Now, since every square can be produced in constant time, the time-complexity for H_2 is $O(n\log n+\mu(F))$ and the number of squares produced is $H_2(P)=\mu(F)$, where $\mu(F)$ is the minimum number of squares needed to cover the set of funnels in P.

Step 3: Partitioning the edges into groups (H_3)
Partition the remaining uncovered edges of P into groups, $g_1,\ldots,g_r$, such that all the edges in a group can be included in a line with the same slope as the edges in the sequence. Partitioning the edges into $O(n)$ groups using an ordinary sorting-algorithm takes $O(n\log n)$ time.

Step 4: Partitioning the groups into sequences (H_4)
Now partition the groups into sequences such that all the edges in a sequence may be included in a non-empty rectangle. For each group g_i we do the following: let l_i be the line that includes all the edges in a group g_i. We check for every edge of P, whether the edge crosses l_i. If some edge crosses the line between two

edges in g_i, the group is partitioned into two smaller groups. If no edge of P crosses l_i then the group is said to be a *sequence*. Continue as described until all the groups are partitioned into sequences. The sequences are denoted $s_1, \ldots, s_p$, and note that a sequence may consist of one single edge. The time-complexity of H_4 is $n \cdot \sum_i O(\log |g_i|) = O(n^2)$, since we have to go through all the edges for every group.

Step 5: Cover the remaining uncovered Voronoi faces (H_5)
This step of the algorithm consists of independently processing each sequence and outputting a set of squares which covers the Voronoi faces induced by the edges in the sequence, and the concave faces induced by the concave vertices of the edges in the sequence. Let E be an arbitrary sequence of P. Rotate the polygon, in such a way that the edges in E are horizontal and at the bottom, and let $e_1, e_2, \ldots, e_k$ be the set of edges in E, ordered from left to right. Recall that every edge, e, of P induces a Voronoi face, $V(e)$, in P, Fig 1a.

We will now produce a set of squares, S, that covers the Voronoi-cells, $V(e_1), \ldots, V(e_k)$, and the faces induced by the concave vertices in E, and then prove that the number of squares in S is at most $24 \times \mu(P)$.

A square in a polygon is said to be *maximal* if it is not contained by any larger square lying in the polygon.

The covering is done in three simple steps.

1. Place the largest possible square on the sequence with lower left corner at e_1's left endpoint. If the square isn't maximal extend it to the left. Denote this square s_1. If s_1's lower right corner lies between two edges e_i and e_{i+1} then place the largest possible square on the sequence with lower left corner at e_{i+1}'s left endpoint. If the square isn't maximal extend it to the left. This square is denoted s_2. Otherwise, if s_1's lower right corner lies on an edge $e_i \in E$ place the largest possible square on the sequence with lower left corner at s_1's lower right corner, Fig. 2a. If the square isn't maximal extend it to the left. This square is denoted s_3. Continue to cover the edges as described until all the edges in E are entirely included by $S = \{s_1, \ldots, s_q\}$.

2. Let $R(s_i)$, $1 \leq i \leq q$, be the Voronoi region or regions that are induced by the edges in the sequence, such that every point in $R(s_i)$ has a perpendicular projection on s_i's base. If $R(s_1)$ isn't entirely covered by s_1 then let p be that uncovered point in $R(s_1)$ with longest perpendicular projection on s_1's base. Now, place a square s_{q+1} with center in p and with bottom corner at p's projection on s_1's base, Fig. 2a. It is easily seen that s_{q+1} covers the remaining uncovered region of $R(s_1)$ and, that s_{q+1} lies entirely within P. Continue to cover the remaining uncovered regions of $R(s_2), \ldots, R(s_q)$.

3. It remains to cover the faces induced by the concave vertices of P. Let C be a concave vertex of P. Split the concave angle into two equal angles. The two resulting regions, c_1 and c_2, are denoted *concave cells*, Fig. 2b.

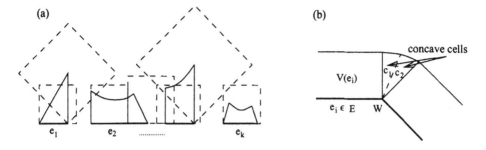

Fig. 2. (a) The Voronoi faces induced by $e_1, \ldots, e_k$ are entirely covered by the squares (dashed) produced in Steps 1-2. (b) The concave cell c_1 belongs to E.

Definition 3. If a concave cell, c, shares a Voronoi edge with any of the Voronoi faces induced by the edges of a sequence, E, then c is said to *belong* to E.

A concave cell can be covered by two squares within P as follows. The Voronoi edge of the cell with one endpoint in the concave vertex W is called the *arm* of the cell. Partition the cell into two subcells by splitting the angle at W into two equal angles. Calculate the point, p_1 (p_2) in each subcell with largest distance to W. Construct two squares such that they both have one corner in W and center in p_1, respectively p_2. It is easily seen that these two squares cover the cell and, according to the definition of the medial axis, they lie entirely inside the polygon.

Since the edges in E are horizontal and at the bottom, it holds that every concave cell that belongs to E lies entirely above, (i.e. has larger y-coordinates than), the edges in E. Thus a square that includes two or more edges of E covers all the concave cells that belong to E, between these edges. Now, cover every uncovered cell that belongs to E.

Let OPT be a set of $\mu(P)$ squares that covers P and let $OPT(E)$ be a minimum set of squares that covers $V(e_1), \ldots, V(e_k)$. It is obvious that $|OPT(E)|$ is greater than or equal to q (the number of squares produced in step 5.1).

There are at most two uncovered concave cells for every square produced in step 5.1, thus the number of squares in S is at most 6 (=2 squares + (2 concave cells $\times$ 2 squares)) times greater than the number of squares in $OPT(E)$. A square in OPT may include edges or part of edges at all its four sides, and since the squares produced by H_5 cover one sequence at a time it's possible that every square in S only includes edges on one side. Thus the number of squares produced in this step is at most 2 squares +(2 concave cells $\times$2 squares)$\times$4 sides=24 times the number of squares in OPT.

Constructing a square takes $O(n)$ time and there are at most $O(n)$ squares, according to Observation 2, constructed in this step, since the funnels are covered by H_2. Hence, the time-complexity for H_5 is $O(n^2)$.

Steps 1 and 3 take $O(n \log n)$ time, Step 2 takes $O(n \log n + \mu(P))$ time and, finally, Steps 4 and 5 take $O(n^2)$ time. Thus the total time-complexity for H is $O(n^2 + \mu(P))$.

Summarizing the main result of this section we obtain the following theorem.

Theorem 4. *For any polygon P, without any acute interior angles, with n vertices, Algorithm H produces a covering of P with at most $24 \times \mu(P)$ squares in $O(n^2 + \mu(P))$ time.*

2.1 Improvements

In this section we will show how to improve the approximation factor, from 24 to 14, by producing each square within P that includes edges or parts of edges at three or four of its sides.

Definition 5. A square r is said to be a 3-square in P iff r includes edges or parts of edges (of length > 0) of P on at least three of its sides.

The new algorithm is denoted H', and it's an extended version of H. The only difference between Steps 1-4 of H and H' is that an extra step is added between Step 3 and 4 in H', which produces all 3-squares and then marks the edges in the sequences included in these squares.

To find all 3-squares in P we first group the edges in P into a minimum number of groups, $S_1, \ldots, S_p$, such that the slope of the edges in S_i, $1 \le i \le p$, differ by exactly a multiple of $90°$. This grouping can easily be done in $O(n \log n)$ time by first sorting the slopes, and then scanning the sorted list.

Next, for each group we find its 3-squares by first rotating the polygon (or the coordinate system), such that the edges in the group are vertical or horizontal, and then constructing the generalized Voronoi diagram for these edges in the L_∞-metric, Fig. 3. Every vertex and isothetic edge in the Voronoi diagram corresponds to an isothetic rectangle. For every Voronoi vertex the corresponding (isothetic) rectangle is the maximal square, whose center lies at the vertex. Thus, every Voronoi vertex connecting three or four Voronoi edges corresponds to a square, whose center lies at the vertex.

Then each square is examined to decide whether it is a 3-square within P. The total number of squares examined according to the above procedure is $O(n)$, and for each one of them it takes $O(n)$ time to decide whether it is a 3-square within P. Thus, the total time to find all valid 3-squares in P is $O(n^2)$.

The edges or parts of edges that are included in the boundaries of the 3-squares are marked in the groups produced in H'_3 before H'_4. Now, all the 3-squares in an optimal covering are produced and we will now prove that the total number of squares produced by H' is at most $14 \times \mu(P)$.

To improve the approximation factor we have to modify the step of H' corresponding to H_5 (Section 2), denoted H'_5.

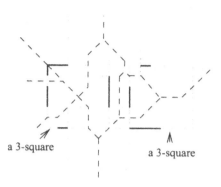

a 3-square

a 3-square

Fig. 3. The Voronoi diagram (dashed) in L_∞-metric.

Step H'_5: Let $\mathcal{E}$ be an arbitrary sequence of P, and let R be the set of 3-squares produced by H', such that every square in R includes edges or parts of edges of $\mathcal{E}$. Rotate P and all the produced squares, such that $\mathcal{E}$ is horizontal and at the bottom. Order the squares in R from left to right $r_1, \ldots, r_s$. Let E_1 be the subset of $\mathcal{E}$ to the left of r_1, let E_{s+1} be the subset of $\mathcal{E}$ to the right of r_s and let E_i, $2 \leq i \leq s$, be the subset of $\mathcal{E}$ between r_{i-1} and r_i. The subset E_i, $1 \leq i \leq s+1$, is denoted a *subsequence* of $\mathcal{E}$, Fig. 4. Note that a subsequence may be an empty set. The edges or parts of edges in E_i are denoted $e_{i_1}, \ldots, e_{i_{q_i}}$ (ordered from left to right).

Now, for every subsequence of $\mathcal{E}$ we will cover the cells that are *associated* to E_i (to be defined below), $1 \leq i \leq s+1$, in three different ways. Then we will select the covering which produces the smallest number of squares. The three different coverings are computed as follows:

Covering 1. Place the largest possible square on the sequence with lower left corner at e_{i_1}'s left endpoint. If the square isn't maximal extend it to the left. Denote this square $s_1^1(E_i)$. If $s_1^1(E_i)$'s lower right corner lies between two edges e_{i_j} and $e_{i_{j+1}}$, $1 \leq j < q_i$, then place the largest possible square on the subsequence with lower left corner at $e_{i_{j+1}}$'s left endpoint. If the square isn't maximal extend it to the left. This square is denoted $s_2^1(E_i)$. Otherwise, if $s_1^1(E_i)$'s lower right corner lies on an edge e_{i_j} place the largest possible square on the sequence with lower left corner at $s_1^1(E_i)$'s lower right corner, Fig. 2a. If the square isn't maximal extend it to the left. This square is then denoted $s_2^1(E_i)$. Continue to cover the edges as described until all the edges in E_i are entirely included by $SEQ'_1(E_i) = \{s_1^1(E_i), \ldots, s_{k_1}^1(E_i)\}$. Note that the number of squares in $SEQ'_1(E_i)$ is equal to the minimum number of squares needed to include the edges or parts of edges in E_i.

Covering 2. Instead of placing the first square on e_{i_1}'s left endpoint, we place the largest possible square, s_1^2, on the sequence with lower left corner at r_{i-1}'s bottom right corner. Continue as described in the previous paragraph until

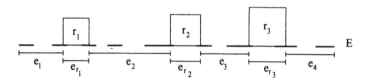

Fig. 4. A sequence E divided into subsequences.

all the edges in E_i are entirely included by $SEQ'_2(E_i)=\{s_1^2(E_i),\ldots,s_{k_2}^2(E_i)\}$. Note that this covering is not done for E_1.

Covering 3. In the last covering we cover the subsequence from right to left, thus we do exactly as in the previous covering, but instead we place the largest possible square on the subsequence with lower right corner at r_i's bottom left corner. Now, continue as described in Covering 1 (but now from right to left) until all the edges in E_i are entirely included by $SEQ'_3(E_i)=\{s_1^3(E_i),\ldots,s_{k_3}^3(E_i)\}$. Note that this covering is not done for E_{s+1}.

The edges in E_i are now included in three different coverings. Before we compare the three different coverings, we have to cover the remaining uncovered regions of the cells that are *associated* to the subsequence. We define them below.

Let E_i be an arbitrary subsequence of $\mathcal{E}$, let $C(E_i)=\{c_1(E_i),\ldots,c_{p_i}(E_i)\}$ be the concave cells that belong (see Definition 3) to E_i ordered from left to right and, let e_{r_i} be the edges or parts of edges in $\mathcal{E}$ included in r_i. The left and right endpoints of e_{r_i} may be concave vertices that induce concave cells. If the left endpoint of e_{r_i} induces two concave cells then the concave cell entirely above $\mathcal{E}$ is denoted $c_{p_{i-1}+1}(E_{i-1})$; this cell is said to be *associated* to E_{i-1}. If the right endpoint of e_{r_i} induces two concave cells then the concave cell entirely above $\mathcal{E}$, $c_0(E_i)$, is *associated* to E_i. The concave cells $c_1(E_i),\ldots,c_{q_i}(E_i)$ and $c_0(E_i)$ and/or $c_{q_i+1}(E_i)$, if they exist, plus the Voronoi cells induced by the edges in E_i are said to be *associated* to E_i.

The remaining uncovered cells associated to E_i are covered as described in Steps 2 and 3 of H_5. The three resulting sets of squares are denoted $SEQ_1(E_i)$, $SEQ_2(E_i)$ and $SEQ_3(E_i)$. Now all the cells associated to the subsequence are covered by three different coverings, and we can easily compare the number of squares in each of them, and then select a covering which produces the smallest number of squares. The smallest set is denoted $SEQ(E_i)$. This is done for all the subsequences of P.

Definition 6. Let r be an arbitrary 3-square in P. If r overlaps with edges of P on all its four sides then let any of the four sides be r's *base*. Otherwise, one side of r doesn't overlap with any edges of P, and the opposite side of r is then called the *base* of r.

There might still be uncovered concave cells that belong to the edges in $\mathcal{E}$. Each such concave cell has to be induced by some concave vertex lying on the base of a 3-square. Let r be an arbitrary 3-square in R that includes edges or parts of edges on three of its sides. We may assume that r's base is horizontal and at the bottom. There may be uncovered concave cells, belonging to the edges or parts of edges lying on r's base, that sticks out above the top of r. To cover these uncovered cells let p_r be the point (with a perpendicular projection on r's base) in the uncovered cells with largest distance to r's base and let p'_r be the perpendicular projection of p_r on r's base. Place a maximal square r' with bottom corner on p'_r and center in p_r. It is easily seen that all the remaining uncovered concave cells that belongs to the edges or parts of edges lying on r's base is covered by r'. We say that r' is the *companion* of r.

In the full version [6] we prove that by using this improved algorithm, we will obtain a covering which is within a factor fourteen of the optimal.

3   A Linear-Time Algorithm

In this section we will show a covering algorithm, H, that covers an arbitrary polygon P with at most $11n+\mu(P)$ squares in linear time, provided that the medial axis of the polygon is given. The algorithm works by partitioning the polygon P into cells which are then covered pairwise by squares lying within P. The first step of our algorithm is achieved by drawing the Voronoi diagram, $V(P)$, described in Section 2, Step 1. The diagram can be computed in $O(n \log n)$ time [9] for an arbitrary polygon and in linear time for a simple polygon [4]. The following has been shown:

Observation 7. *The number of edges in $V(P)$ is at most $3n$, where n is the number of vertices in the polygon P.*

Proof. Omitted in this version, see [6].

Now, for every Voronoi face f, induced by an edge of P, let g be the segment of P which induces the face. The next step of our algorithm is achieved by partitioning each face into cells with only three or four vertices each, by drawing from every vertex of the face a segment connecting the vertex with its perpendicular projection on g.

Definition 8. We call the edge of each cell which is at the boundary of P the *base* of the cell. We call the segments which are perpendicular to its base the *sides* of the cell, the edge opposite to its base is called the *top* of the cell, and finally, two cells sharing a Voronoi edge, e, are called a *proper pair of cells*.

Since there are at most $3n$ edges in $V(P)$, it remains to show that every proper pair of cells can be covered by a constant number of squares or that the number of squares produced to cover a (proper) pair of cells is optimal, to obtain the promised result. (The idea of processing Voronoi cells separately was also proposed in [11].) We will in the following section show that every pair of cells may be covered by at most four squares, except for a total of at most $\mu(P)$ additional squares, thus producing at most $4 \cdot 3n + \mu(P) = 12n + \mu(P)$ squares.

3.1 Covering the cells

The rest of the algorithm consists of independently processing each (proper) pair of cells and outputting a set of squares which covers them. Each pair is processed in time linear with respect to the number of squares produced in this step. The total number of edges is less than $3n$ and therefore the overall time performance of this step is $O(n+s)$, where s is the number of squares produced in this step. We will now describe how each pair of cells is processed.

There are four major cases: (1) the Voronoi edge between the two cells is a paraboloid, (2) when the two opposite cells are trapezoids, (3) when the two opposite cells are triangles, and (4) when the two opposite cells are both induced by concave vertices.

In the continuation of this text we will denote the cells in a proper pair C_l and C_t. If any of the cells are induced by an edge of P then we may assume without loss of generality that C_l is induced by an edge of P, C_l lies below C_t and that the base of C_l is horizontal and at the bottom.

Case 1. The Voronoi edge between C_l and C_t is a paraboloid, i.e. C_l is induced by an edge and C_t is induced by a concave vertex.

In this case the cells are covered by at most three squares. The proof is omitted in this version, see [6].

Case 2. C_l and C_t are both trapezoids.

In this case the cells are covered by at most four squares, unless they form a funnel. Let A, B, C and D be the corners of C_l in a counter-clockwise order, where A is the left endpoint of C_l's base, and let A' and B' be the left, respectively right, endpoint of C_t's base and thus, the symmetrical images of A and B, Fig. 5. We may assume w.l.o.g. that $|BC| \geq |AD|$. Let α be the interior angle $\angle ADC$. There are two subcases:

Subcase 2.1 $\alpha \geq 135°$ or $|AB| < |AD|$.
Place a square s_1 on AB, such that a side of s_1 coincides with AB, and a square s_2 with center at C and bottom corner at B. These two squares lie entirely within P and cover C_l, Fig. 5a. Cover C_t in the same way as C_l.

Subcase 2.2 $\alpha < 135°$ and $|AD| \leq |AB|$. ("funnel")
In this case we cover the funnel in the same way as described in Section 2, Step 2. Since every edge has to be entirely covered in any optimal partition, it is easily seen that the total number of squares produced in this substep is not greater than $4k+\mu(P)$, where k is the number of pairs of cells belonging to this subcase.

Note that the number of squares needed to cover a funnel can be calculated easily in constant time.

Case 3. The two opposite cells are both triangles.

In this case the cells are covered by at most three squares. Let A, B and C be the corners of C_l in a counter-clockwise order, where A is the left endpoint of C_l's base, and let B' be the left endpoint of C_t's base. Note that $45° \leq \angle BAC < 90°$.

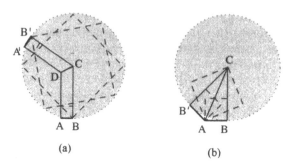

Fig. 5. Case 2.1 and 3.

Place two squares s_1 and s_2, such that s_1 coincides with AB and s_2 coincides with AB'. Finally, place a square, s_3, with bottom corner on A and with upper diagonal corner on C. It is easily seen that s_3 covers the remaining uncovered area, since the interior angle $\angle(BCA)$ is less than $90°$ and the crossing between s_3's sides and BC and $B'C$ are never above s_1's upper right corner respectively s_2's upper left corner. These three squares lie entirely within P, since we know that there exists a circle within P with center in C and perimeter on B and B', according to the definition of the medial axis.

Case 4. The two opposite cells are both induced by a concave vertex.

In the full version [6], we show that in this case the cells are covered by at most three squares.

From Observation 7 we have that P is partitioned into at most $3n$ pair of Voronoi faces, and in Section 3.1 we have shown that every pair can be covered by at most four squares or, subcase 3.2, by an optimal number of squares. Hence, H will produce at most $12n+\mu(P)$ squares to cover P in $O(n\log n+\mu(P))$ time.

As we can see there is just one subcase, namely 2.1, where we may need four squares to cover a pair of cells. We may improve our upper bound by noting that the maximum number of trapezoidal pair of cells is at most $2n$. By using this result it is not difficult to show that the algorithm will produce less than $4 \times 2n + 3 \times n + \mu(P)$ squares to cover P, thus

$$H(P) \leq 11n + \mu(P).$$

3.2 A Tight Lower Bound on Optimal Coverings for Hole-Free Polygons

Theorem 9. *For all integers $n \geq 4$ and for every hole-free polygon P with n vertices it holds that $\mu(P) = \Omega(\frac{n}{\alpha(n)})$.*

Proof. Let $G = \{l_1, \ldots, l_{4\mu(P)}\}$ be a collection of m segments in the plane. Let P be the polygon bounded by the outer face in the partition of the plane induced

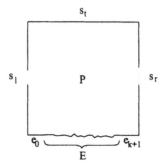

Fig. 6. It is possible to cover P with $O(n/\alpha(n))$ rectangles.

by G. Y_G is a piecewise linear function, whose graph consists of subsegments of the segments l_i, $1 \leq i \leq m$. Hart and Sharir [8] have shown that Y_G consists of $O(m \cdot \alpha(m))$ segments.

The number of edges of P is bounded by $O(m \cdot \alpha(m))$ [8]. Since the total number of segments bounding the squares in any minimum square covering of P is $\leq 4 \cdot \mu(P)$ (four segments per square), we obtain, as above, that $n = O(\mu(P) \cdot \alpha(\mu(P)))$. Thus, $\mu(P) = \Omega(\frac{n}{\alpha(\mu(P))})$ and, hence $\mu(P) = \Omega(\frac{n}{\alpha(n)})$.

Theorem 10. *The lower bound shown in Theorem 9 cannot be generally improved, that is for each n there is a polygon P with $n \geq 10$ vertices such that* $\mu(P) = O(\frac{n}{\alpha(n)})$.

Proof. According to Wiernik [18] there exists a construction of a set M, of m straight-line segments, such that the lower envelope Y of M consists of $\Omega(m \cdot \alpha(m))$ subsegments.

Construct k non-vertical connected edges, $E = \{e_1, \ldots, e_k\}$, such that the edges in E are identical to a lower envelope produced by the m segments in [18]. Let e_i be the i:th left-most edge of E and let d denote the largest vertical distance between two points in E.

Expand E along the x-axis, by multiplying all x-coordinates such that the minimum difference along the x-axis between two incident vertices in E is $10d$.

Let e_0 and e_{k+1} be two horizontal edges of length $10d$, such that e_0 is appended to the left side of e_1 and e_{k+1} is appended to the right side of e_k. Let E' be the set of edges in E plus e_0 and e_{k+1}. Now, construct an upside-down histogram P, with one horizontal upper long-side, s_t, two vertical edges s_l and s_r, of length between $2|s_t| - d$ and $2|s_t|$ such that s_l connects s_t's left endpoint with e_0's left endpoint and s_r connects s_t's right endpoint with e_{k+1}'s right endpoint. The length of s_l and s_r guarantees that a square which includes an edge in E can't include any point on s_t. (See Fig. 6.)

In the full version [6] we prove that it is possible to cover P with $O(\frac{n}{\alpha(n)})$ squares.

Summarizing the main result of this section we obtain the following theorem.

Theorem 11. *For any polygon P with n vertices, Algorithm H produces a square covering of P with at most $11n + \mu(P)$ squares in $O(n \log n + \mu(P))$ time. In the hole-free case the algorithm runs in linear time and produces a cover which is within an $O(\alpha(n))$ approximation factor of the optimal. The number of squares needed to cover P by using the algorithm H can be calculated in $O(n \log n)$ time ($O(n)$ time in the hole-free case).*

The algorithm presented above can easily be parallelized since every cell can be processed in constant or optimal time, thus proposing a natural parallel algorithm. Rajasekaran *et al.* [16] showed in 1994 that the medial axis of a polygon P can be constructed in parallel by $O(n)$ processors in $O(\log n)$ time in the CRCW PRAM model by using random sampling. Hence, our algorithm can easily be parallelized by first constructing the medial axis of P and then covering the cells independently. We obtain a parallel algorithm which runs in $O(\log n)$ randomized time using $O(\max[\mu(P), n])$ processors in the CRCW PRAM model.

4 Fat Rectangular Coverings and Similar Problems

We may use the two algorithms proposed in this paper for covering polygons with other types of polygons, denoted *q-polygons*, provided that the following two properties hold. Let k_1 and k_2 be two arbitrary, but fixed, integers.

1. A q-polygon is coverable by k_1 squares, and
2. A square is coverable by k_2 q-polygons.

q-polygons that fulfill these two properties are, for example, rectangles having bounded aspect ratio.

Now, let $\mu^*(P)$ denote the minimum number of q-polygons needed to cover an arbitrary polygon P. We have that $\mu(P) \leq k_1 \cdot \mu^*(P)$. From the above properties it easily seen that by using the two algorithms proposed in this paper, for the q-polygon cover problem, we will get the following two results:

- An algorithm corresponding to the one presented in section 2 will produce at most $14 \cdot k_1 \cdot k_2 \cdot \mu^*(P)$ q-polygons to cover a polygon P in $O(n^2 + \mu^*(P))$ time.
- An algorithm corresponding to the one presented in section 3 will produce at most $11n \cdot k_2 + k_1 \cdot k_2 \cdot \mu^*(P)$ q-polygons to cover P in $O(n \log n + \mu^*(P))$ time.

References

1. L.J. Aupperle, H.E. Conn, J.M. Keil and J. O'Rourke, *Covering Orthogonal Polygons with Squares*, 26th Annual Allerton Conference on Communication, Control and Computation, 1988.
2. R. Bar-Yehuda and E. Ben-Hanoch, *A Linear-Time Algorithm for Covering Simple Polygons with Similar rectangles*, International Journal of Computational Geometry & Applications, vol. 6, no 1, 1996.

3. B.M. Chazelle, *Computational Geometry and Convexity*, Ph.D. Thesis, Dept. Comp. Sci., Yale University, New Haven, CT, 1979. Carnegie-Mellon Univ. Report CS-80-150.

4. J. Snoeyink, C.A. Wang and F. Chin, *Finding the Medial Axis of a Simple Polygon in Linear-Time*, ISAAC '95, Cairns, Australia, 1995 (LNCS 1006 Springer-Verlag).

5. C. Levcopoulos and J. Gudmundsson, *Close Approximation of Minimum Rectangular Coverings*, FST&TCS-16, Hyderabad, India, 1996 (LNCS, Springer-Verlag).

6. C. Levcopoulos and J. Gudmundsson, *Approximation Algorithms for Covering Polygons with Squares and Similar Problems*, LU-CS-TR:96-181, Dept. of Comp. Sci., Lund University, 1996.

7. A. Hegedüs, *Algorithms for covering polygons by rectangles*, Computer Aided Design, vol. 14, no 5, 1982.

8. S. Hart and M. Sharir, *Nonlinearity of Davenport-Schinzel Sequences and of Generalized Path Compression Schemes*, Tech. Report 84-011, The Eskenasy Institute of Comp. Sci., Tel Aviv University, August 1984.

9. D.G. Kirkpatrick, *Efficient computation of continuous skeletons*, 20th Annual IEEE Symposium on Foundation of Computer Science, 1979.

10. J.M. Keil and J.-R. Sack, *Minimum Decompositions of Polygonal Objects*, Machine Intelligence and Pattern Recognition vol. 2: Computational Geometry, pp. 197-215, Elsevier Science Publishers B.V., 1985.

11. C. Levcopoulos, *A Fast Heuristic for Covering Polygons by Rectangles*, FCT'85, Cottbus, GDR, 1985 (LNCS 199, Springer-Verlag).

12. C. Levcopoulos, *Improved Bounds for Covering General Polygons with Rectangles*, FST&TCS-7, Pune, India, 1987 (LNCS 287, Springer-Verlag).

13. D. Morita, *Finding a Minimal Cover for Binary Images: an Optimal Parallel Algorithm*, Tech. Report No. 88-946, Dept. of Comp. Sci., Cornell University, 1988.

14. J. O'Rourke and K.J. Supowit, *Some NP-hard Polygon Decomposition Problems*, IEEE Transactions on Information Theory, vol. IT-29, pp.181-190, 1983.

15. F.P. Preparata and M.I. Shamos, *Computational Geometry*, New York, Springer-Verlag, 1985.

16. S. Rajasekaran and S. Ramaswami, *Optimal Parallel Randomized Algorithms for the Voronoi Diagram of Line Segments in the Plane and Related Problems*, In Proc. ACM Symposium on Computational Geometry, Stony Brook, New York, 1994.

17. D.S. Scott and S.S. Iyengar, *TID: a Translation Invariant Data Structure for Storing Images*, Comm. of the ACM, vol. 29, no. 5, 1986.

18. A. Wiernik, *Planar Realization of Nonlinear Davenport-Schinzel Sequences by Segments*, 27th IEEE Symposium on Computer Science, 1986.

Greedily Approximating the r-independent Set and k-center Problems on Random Instances

Extended Abstract

Bernd Kreuter and Till Nierhoff*

Institut für Informatik, Humboldt-Universität zu Berlin, D-10099 Berlin, Germany

Abstract. In this paper we analyse the performance of the greedy algorithm for r-independent set on random graphs. We show that for almost all instances

- the greedy algorithm has a performance ratio of $2 + o(1)$,
- the greedy algorithm yields a $1 + o(1)$ approximation of the r-dominating set problem,
- the k-center problem can be solved optimally.

1 Introduction

In the context of average case analysis of algorithms the algorithmic theory of random graphs has attracted considerable interest recently. An excellent overview of this field is provided by the survey article by Frieze and McDiarmid [6]. One of the first algorithms investigated was the greedy algorithm for the independent set problem and the colouring problem by Grimmett and McDiarmid [8].

Here we consider the following generalization of the independent set problem. For a positive integer r, a set of vertices W in an (unweighted) graph G is called *r-independent* if any two of its vertices have distance greater than r, where the distance between two vertices in a graph is the length (i.e. the number of edges) of a shortest path between these vertices. A 1-independent set is just an independent set. For a graph G, let $\alpha_r(G)$ denote the size of a maximum r-independent set in G. Generalizing the results of Grimmett and McDiarmid we show that the greedy algorithm for r-independent set achieves for almost all graphs an approximation with ratio 2, i.e. the output of the greedy algorithm has almost surely at least half the size of a maximum r-independent set. So far even for 1-independent sets, no algorithm is known that finds for almost all instances an independent set of size $1/2 + \varepsilon$ of the size of a maximum independent set. The problem to decide whether such an algorithm exists is mentioned as Research Problem 15 in [6].

Similarly, for an integer r, a set of vertices W in a graph $G = (V, E)$ is called *r-dominating* if any vertex from $V \setminus W$ has distance at most r to some vertex from W. A 1-dominating set is usually called a dominating set. By $\gamma_r(G)$ we denote

---

* Fellow of the graduate school "Algorithmische Diskrete Mathematik", which is supported by the Deutsche Forschungsgemeinschaft, grant GRK 219/2-96

the minimal size of an r-dominating set in G. Closely related to the r-dominating set problem is the k-center problem, where k is a positive integer. Here, the aim is to find a set of k vertices minimizing the maximal distance of a vertex from $V\backslash W$ to the set W. There is a vast literature on these parameters, cf. [9] for a bibliography up to 1990. Chang and Nemhauser [5] proved the $\mathcal{NP}$-completeness of the r-independent set and r-dominating set problems, in the case of $r \geq 2$ even for bipartite graphs.

A maximal r-independent set is also an r-dominating set. In Section 3 we show that the greedy algorithm for r-independent set yields with high probability an r-dominating set of optimal size, apart from lower order terms. This implies that also the dual k-center problem can be solved almost optimally. We show furthermore that an even simpler algorithm than the greedy algorithm, which just randomly selects a vertex set of the right size, yields with high probability an optimal solution to the r-dominating set resp. the k-center problem.

The distribution of r-independent sets and r-dominating sets of a fixed size s was investigated by Burtin [4]. He shows that the number of r-independent sets resp. r-dominating sets of size s is close to a Poisson distribution for a certain range of p.

The model for random instances used here is the so called $G_{n,p}$ model for random graphs where n is the number of vertices and edges are present independently with probability $p = p(n)$. For an introduction to the theory of random graphs cf. [2].

The k-center problem can be stated in many variants, according to the many applications that it has, e.g. the problem of distributing hospitals in a town. The model treated here is a good description of the k-center problem in a network where one has to determine k servers among n nodes in a way that any non-server node does not have to communicate via too many other nodes in order to reach a server. In this setting the graph model reflects the fact that the transmission of messages is usually fast, compared to the time needed to forward messages at intermediate nodes.

Hochbaum [10] investigates another variant of the k-center problem: The input consists of the complete graph on n vertices K_n and edge lengths, which are i.i.d. random variables. The metrics corresponding to that model are quite different from the distance metrics considered in this paper.

2   The greedy algorithm for r-independent sets

Our object of study is the following algorithm:

Procedure GREEDY-r-INDSET

Input: $G = (V, E)$

Initialization: unmark all of V;
               $i \leftarrow 0$;

Iteration: while there are unmarked vertices:
               $i \leftarrow i + 1$;
               choose v_i from the unmarked vertices;
               mark all vertices within distance r of v_i;

Output: $\{v_1, \ldots, v_i\}$

We call $\{v_1, \ldots, v_i\}$ a *greedy r-independent set on G* and denote by $g_r(G)$ its size. We analyse the asymptotic behaviour of the algorithm under the assumption that the input G is distributed according to the $G_{n,p}$-model of random graphs, where $p = p(n) \in (0,1)$. We write $q := 1 - p$ and $b := -\log q$ for short; log denotes the natural logarithm. Then $p \leq b$ and $p \geq (1 - o(1))b$ for $p = o(1)$. The latter condition is fulfilled in most parts of this paper, since for $r \geq 2$ our assertions take effect only for $p \leq n^{-1+1/r}(\log n)^{1/r}$.

We say that an event for graphs holds *with high probability* (abbreviated by *whp.*) if the probability that it holds in $G_{n,p}$ tends to 1 when n tends to infinity.

For two functions $f, g : \mathbb{N} \to \mathbb{R}$, we write $f \approx g$ iff for every $\varepsilon > 0$ there exists an n_0 such that for any $n \geq n_0$

$$\lfloor (1 - \varepsilon)g(n) \rfloor \leq f(n) \leq \lceil (1 + \varepsilon)g(n) \rceil.$$

The brackets $\lfloor . \rfloor$ and $\lceil . \rceil$ are of course only important if $g(n) = \mathcal{O}(1)$. We introduce them here in order to avoid a separate treatment of this case. To simplify the presentation we omit the brackets throughout this paper. Moreover the notation $f = \omega(g)$ means that f/g tends to infinity as n tends to infinity.

Our main result can now be stated as follows:

Theorem 1. *Let $r \geq 1$ be an integer and $p = \omega(\log n/n)$. Then the greedy algorithm outputs whp. a solution of size*

$$g_r(G_{n,p}) \approx \frac{r \log np}{(np)^{r-1}b},$$

where $b = -\log(1 - p)$.

The restriction on p with $p = \omega(\log n/n)$ is due to technical reasons in the proof, namely the applicability of the Chernoff-bounds in Lemma 3. It might be considered natural, because $p_0 = \log n/n$ is the threshold function for connectivity.

Theorem 1 is complemented by the next theorem which gives an estimate for the size of an r-independent set in $G_{n,p}$:

Theorem 2. *Let $r \geq 1$ be an integer and $p = \omega(1/n)$. Then whp.*

$$\alpha_r(G_{n,p}) \leq (1 + o(1)) \frac{2r \log np}{(np)^{r-1} b}.$$

Thus, GREEDY-r-INDSET yields whp. an r-independent set that has at least half of the size of a maximum r-independent set.

For the range of $p = p(r, n)$ where $s = \frac{2r \log np}{(np)^{r-1} b} = \mathcal{O}(1)$, Burtin [4] showed moreover that the number of r-independent sets of size s is asymptotically Poisson distributed. For $r = 1$, Frieze [7] showed that the bound on $\alpha_1(G_{n,p})$ is tight and also determines lower order terms.

2.1 Notation and tools

We first introduce some notation: Let

$$B(G, v, r) := \{w \in V(G) : d_G(v, w) \leq r\},$$

the ball of radius r around v in the graph G and

$$S(G, v, r) := \{w \in V(G) : d_G(v, w) = r\},$$

the sphere of radius r around v in G. Let $b(G, v, r) = |B(G, v, r)|$ and $s(G, v, r) = |S(G, v, r)|$. For $W \subseteq V(G)$ let $B(G, W, r) := \bigcup_{w \in W} B(G, w, r)$ and similarly for b, S and s. We sometimes omit the G-parameter if its choice is clear from the context. For a graph G and a subset W of its vertices, we write $G[W]$ for the graph induced on W.

We write $X \sim D$, if the random variable X has distribution D, and $X \preceq D$ if for $Y \sim D$ and any constant c,

$$Pr[X \leq c] \geq Pr[Y \leq c].$$

For two random variables X and Y, $(X|Y = y)$ denotes the conditional distribution of X, given $Y = y$. Observe that if $Y \sim Bi(n, p_1)$ and $(X|Y = y) \sim Bi(y, p_2)$ then

$$X \sim Bi(n, p_1 p_2). \tag{1}$$

The following tool by Burtin (Lemma 1 in [3]) will play a central role in our analysis:

Lemma 3. *For $p \leq n^{-1+1/r} (\log n)^{1/r}$, let $c_n := (np/\log n)^{1/4}$. Suppose that $c_n = \omega(1)$. Then for every $i = 1, \ldots, r-1$,*

$$Pr\left[|s(v, i) - (np)^i| \leq (np)^i / c_n\right] \geq 1 - n^{-c_n}.$$

The proof is straightforward by repeated application of the Chernoff-bounds and we omit it here. (In [3] the lemma is stated with b instead of p. We furthermore set $c'_n = c_n^4$ and $\omega_n := c_n^2$. Then the condition $(np)^{r-1} \leq n/c_n^2$ follows from the bound on p.)

2.2 Random processes

In order to analyse the greedy algorithm we define some random processes. Let Z_i be the number of unmarked vertices at step i, i.e.

$$Z_i = Z_i(G) = |V(G) \backslash \bigcup_{j=1}^{i} B(G, v_i, r)|.$$

Then $Z_0 = n$. In these terms, $g_r = \min\{i : Z_i = 0\}$. For completeness, we define $Z_i = 0$ for $g_r + 1 \leq i \leq n$.

Furthermore let

$$\mathcal{N}_i = V(G) \backslash \bigcup_{j=1}^{i} B(v_i, r-1) \quad \text{and} \quad N_i = |\mathcal{N}_i|.$$

A central point for the analysis is that edges with both endvertices in $\mathcal{N}_i$ are "unexposed" at stage i:

Call a potential edge $\{v, w\}$ *exposed* at some step if the greedy algorithm has checked whether or not $\{v, w\}$ is in the graph. Assume that at some moment, for some set of vertices W, no edge with both endvertices in W is exposed. Then, since the edges in the $G_{n,p}$ model are selected independently, the graph induced on W is still completely random, i.e. has distribution $G_{\ell,p}$, where $\ell = |W|$. It is easy to check that actually, at stage i, all edges spanned by $W = \mathcal{N}_i$ are unexposed.

The main observation in the proof of the upper bound will be that N_i stays close to n with high probability until the end of the algorithm. This means that in each step we deal more or less with a random graph on n vertices and each v_i expands like a vertex in a random graph on n vertices.

2.3 Proof of Theorem 1

The lower bound follows immediately from the fact that $g_r(G_{n,p}) \geq \gamma_r(G_{n,p})$ and Theorem 4 proven in Section 3.1.

To prove the upper bound, let $0 < \varepsilon < 1$ be given.

Define $k = k(n, p, r) = (1 + \varepsilon) \frac{r \log np}{(np)^r - 1 b}$. Then we have to show that whp. $g_r \leq k$.

Consider first the case that $p \geq p_0 = n^{-1+1/r}(\log n)^{1/r}$. Then $g_r(G_{n,p}) \preceq g_r(G_{n,p_0})$ and it is easily verified that $k(n, r, p_0) \leq 1 + o(1)$. Hence it is enough to prove the theorem for the case $p \leq n^{-1+1/r}(\log n)^{1/r}$, which we will assume to hold from now on.

The proof proceeds as follows: By induction, we show that Z_i is stochastically dominated by some binomial distribution. In the induction step we need to condition on the event that the vertices v_i expand nearly as expected. Applying Lemma 3, we show that these events hold with high probability. Using the estimate on Z_i, we finally show that whp. $g_r \leq k$.

Precisely we are going to show that for any $0 \leq i \leq k$ whp.

$$Z_i \preceq Bi(n, q^{i(1-\varepsilon/4)(np)^{r-1}}). \tag{2}$$

(2) is certainly true for $i = 0$. Using observation (1) it is enough to show that, for $1 \leq i \leq k$, with probability say $1 - 2n^{-2}$,

$$(Z_i | Z_{i-1} = z) \preceq Bi(z, q^{(1-\varepsilon/4)(np)^{r-1}}). \tag{3}$$

To prove the induction step (3) let $i \leq k$ and let $\mathcal{A}_i$ be the event that for $j = 1, \ldots i-1$,

$$b(G, v_j, r-1) \leq 2(np)^{r-1}.$$

Recall from Lemma 3 that $c_n^4 = \frac{np}{\log n}$. By Lemma 3, $\mathcal{A}_i$ holds with probability at least $1 - in^{-c_n} \geq 1 - n^{-2}$, provided n is large enough. If $\mathcal{A}_i$ holds then

$$N_{i-1} \geq n - i2(np)^{r-1} \geq n - \frac{2(1+\varepsilon)r \log np}{p}$$

$$\geq n - \frac{2(1+\varepsilon)r}{c_n^4} n \geq \left(1 - \frac{\varepsilon}{5(r-1)}\right) n$$

provided n is large enough.

Employing that $G[\mathcal{N}_{i-1}]$ is independent of the exposed graph we conclude further by Lemma 3 that

$$s(G[\mathcal{N}_{i-1}], v_i, r-1) \geq (1 - 1/c_{N_{i-1}})(N_{i-1}p)^{r-1} \geq (1 - \varepsilon/4)(np)^{r-1} \tag{4}$$

with probability at least $1 - N_{i-1}^{-c_{N_{i-1}}} \geq 1 - n^{-2}$ (if $\mathcal{A}_i$ holds and n is large enough). Let $\mathcal{B}_i$ be the event that $\mathcal{A}_i$ and (4) both hold. Then

$$Pr[\mathcal{B}_i] \geq 1 - 2n^{-2} \tag{5}$$

Assume now that the algorithm explores the r-neighbourhood of v_i by breadth-first-search in step i. Then, after the exploration of $B(v_i, r-1)$, an unmarked vertex remains unmarked, if there is no edge connecting it to $S(v_i, r-1)$. These potential edges are still unexposed and the edges corresponding to different unmarked vertices are disjoint. Hence, vertices stay unmarked independently, with probability $q^{s(v_i, r-1)}$. Since we condition on $\mathcal{B}_i$, this may by (4) be estimated as

$$q^{s(v_i, r-1)} \leq q^{s(G[\mathcal{N}_{i-1}], v_i, r-1)} \leq q^{(1-\varepsilon/4)(np)^{r-1}}.$$

Furthermore all vertices from $B(v_i, r-1)$ are marked (with probabilty 1). Thus (3) is true if $\mathcal{B}_i$ holds i.e. with probability at least $1 - 2n^{-2}$ (cf. (5)). This concludes the proof of (2).

Since $Z_i \leq Z_{i-1} - 1$ for $i \leq g_r$, by backwards induction $g_r \leq \ell + Z_\ell$ holds for any ℓ.

Choose $\ell = (1 + \varepsilon/2)\frac{r \log np}{(np)^{r-1}b}$. Then as $\varepsilon > 0$,

$$E[Z_\ell] \leq nq^{\ell(1-\varepsilon/4)(np)^{r-1}} = n\exp\{-(1 - \varepsilon/4)(1 + \varepsilon/2)r \log np\} \leq n(np)^{-r}. \tag{6}$$

Therefore, by Markov's inequality,

$$Pr\left[Z_\ell \geq \frac{\varepsilon \log np}{2(np)^{r-1}b}\right] \leq \frac{2b}{p\varepsilon \log np} = o(1),$$

so that whp. $g_r \leq \ell + \frac{\varepsilon \log np}{2(np)^{r-1}b} \leq k$, completing the proof of Theorem 1. $\qquad \square$

2.4 Proof of Theorem 2

The case $r = 1$ is proven in [7], so let now $r \geq 2$. As pointed out in the beginning of the section, we may assume then that $p = o(1)$. Let $W \subset V(G_{n,p})$ be fixed with $|W| = (1 + \varepsilon)\frac{2r \log np}{(np)^{r-1}b} =: k$. We will show that

$$Pr[W \text{ is } r\text{-independent}] \leq (np)^{-rk} \tag{7}$$

for n large enough. Then, by Markov's inequality,

$$Pr[\alpha_r(G_{n,p}) \geq k] \leq \frac{\binom{n}{k}}{(np)^{rk}} \leq \left(\frac{ne}{k(np)^r}\right)^k = \left(\frac{eb}{(1+\varepsilon)2rp \log np}\right)^k = o(1)$$

and thus whp. $G_{n,p}$ does not contain an r-independent set of size k.

To show (7), let I' be the set of all potential paths with both endvertices in W and length at most r. Denote by I the subset of those paths in I' which have length exactly r. Let B_i, $i \in I'$, be the event that all edges of path i are present. Then

$$Pr[W \text{ is } r\text{-independent}] = Pr\left[\bigwedge_{i \in I'} \bar{B}_i\right] \leq Pr\left[\bigwedge_{i \in I} \bar{B}_i\right].$$

Let

$$\mu := \sum_{i \in I} Pr[B_i] = |I|p^r = \binom{k}{2} \cdot (n-2)\ldots(n-r) \cdot p^r$$

and observe that for n (and thus k) large enough, this can be bound by

$$\mu \geq (1-\eta)\frac{k^2}{2}(1-\eta)(np)^{r-1}b \geq (1+\varepsilon/2)k \log(np)^r,$$

where $\eta > 0$ is chosen such that $(1-\eta)^2(1+\varepsilon) \geq (1+\varepsilon/2)$. Thus, if the B_i were independent then $Pr[\bigwedge_{i \in I} \bar{B}_i]$ would be

$$\prod_{i \in I}(1 - Pr[B_i]) \leq exp(-\mu) \leq np^{-(1+\varepsilon/2)rk}.$$

Actually, the B_i are not independent, so we apply Janson's inequality [1, p.96], which reads here

$$Pr\left[\bigwedge_{i \in I} \bar{B}_i\right] \leq \exp\left(-\mu + \frac{\Delta}{2(1-p^r)}\right),$$

where

$$\Delta = \sum_{i \cap j \neq \emptyset} Pr[B_i \wedge B_j]$$

and "$i \cap j \neq \emptyset$" means that paths i and j share an edge.

It is straightforward (though somewhat tedious) to show that $\Delta = o(\mu)$. Since $p = o(1)$, we may assume that $1 - \Delta/(2(1 - p^r)\mu) \geq 1/(1 + \varepsilon/2)$ for n large enough. This shows (7). $\square$

3 Domination problems

3.1 The r-dominating set problem

In the introduction we defined r-dominating sets and $\gamma_r(G)$.

Theorem 4. *Let $r \geq 1$ be an integer and $p = p(n) = \omega(\log n/n)$. Then whp.*

$$\gamma_r(G_{n,p}) \geq (1 - o(1)) \frac{r \log np}{(np)^{r-1} b}.$$

Proof. For $p > n^{-1+1/r}(\log n)^{1/r}$ and n large enough, the right hand side is less than 1 and the statement is therefore certainly true.

For $p \leq n^{-1+1/r}(\log n)^{1/r}$, we apply Lemma 3. Let $0 < \varepsilon < 1$, $k = (1 - \varepsilon) \frac{r \log np}{(np)^{r-1} b}$ and recall that $c_n = (np/\log n)^{1/4}$. By Lemma 3 whp. $s(v, i) \leq (1 + 1/c_n)(np)^i$ for every $i = 1, \ldots, r - 1$ and every vertex $v \in V(G_{n,p})$. In the sequel we condition on this event.

Let $W \subset V(G_{n,p})$ be a fixed set of size k. Then $s(W, r-1) \leq (1+1/c_n)k(np)^{r-1}$. Similarly, we may assume that $b(W, r-1) \leq n/2$.

If W is an r-dominating set then every vertex from $V(G_{n,p}) \backslash B(W, r-1)$ must have a neighbour in $S(W, r-1)$. As $|V(G_{n,p}) \backslash B(W, r-1)| \geq n/2$ the probability of this event is at most

$$\left(1 - q^{(1+1/c_n)k(np)^{r-1}}\right)^{n/2} = \left(1 - \exp\{-b(1 + 1/c_n)k(np)^{r-1}\}\right)^{n/2}$$

$$= \left(1 - (np)^{-(1+1/c_n)(1-\varepsilon)r}\right)^{n/2}$$

$$\leq \exp(-(n/2)(np)^{-(1-\varepsilon/2)r})$$

$$\leq \exp(-kr \log np),$$

where the last inequality follows from the fact that $np = \omega(1)$.

Therefore the expected number of r-dominating sets of size k is at most

$$\binom{n}{k} \exp(-kr \log np) \leq \left(\frac{en}{k}\right)^k (np)^{-rk} = \left(\frac{eb}{(1-\varepsilon)rp \log np}\right)^k = o(1)$$

and by Markov's inequality whp. there is no r-dominating set of size k in $G_{n,p}$. $\square$

Applying that $\gamma_r(G) \leq g_r(G)$ for any graph G, Theorem 4 and Theorem 1 imply the following corollary, which is proven in [11] for $r = 1$.

Corollary 5. *Let $r \geq 1$ be an integer and $p = p(n)$ with $p = \omega(\log n/n)$. Then the greedy algorithm determines whp. an r-dominating set of size*

$$g_r(G_{n,p}) \approx \gamma_r(G_{n,p}) \approx \frac{r \log np}{(np)^{r-1}b}.$$

□

An r-independent dominating set in a graph G is a set of vertices from G that is r-dominating and r-independent or, equivalently, maximal r-independent. The minimum cardinality of such a set is called the r-independent domination number and is denoted by $\iota_r(G)$. $\gamma_r(G)$ and $\iota_r(G)$ can differ by a lot; consider e.g. the graph G on the vertex set $A \cup B \cup \{v, w\}$, where v is adjacent to w and all vertices from A, w is adjacent to v and all vertices from B and there are no further edges. Then $\gamma_r(G) = 2$ but $\iota_r(G) = \min\{|A|, |B|\} + 1$. However, our results imply that these numbers are whp. almost the same.

Corollary 6. *Let $r \geq 1$ be a positive integer and $p = p(n) = \omega(\log n/n)$. Then whp.*

$$\gamma_r(G_{n,p}) \approx \iota_r(G_{n,p}).$$

□

We now discuss an algorithm that is even simpler than the greedy algorithm. RANDOMSELECT-r-DOMSET randomly selects a vertex set W of suitable size and outputs W and all vertices that have distance greater than r of W.

Procedure RANDOMSELECT-r-DOMSET

Input: G, $\varepsilon > 0$
              Pick an arbitrary set $W \subset V$ of size $(1 + \varepsilon/2)\frac{r \log np}{(np)^{r-1}b}$;
              $U \leftarrow V \backslash B(W, r)$;
Output: $W \cup U$

Corollary 7. *Let $r \geq 1$ be an integer, $\varepsilon > 0$ and $p = p(n)$ with $p = \omega(\log n/n)$. Then RANDOMSELECT-r-DOMSET whp. outputs an r-dominating set of size at most $(1 + \varepsilon)\frac{r \log np}{(np)^{r-1}b}$.*

Proof. The correctness of the algorithm is obvious.

As in the proof of the upper bound in Section 2.3 it can be shown (cf. (4)) that whp. $s(W, r-1) \geq (1 - \varepsilon/4)|W|(np)^{r-1}$. Hence the number of vertices in U can be bounded by a binomial distribution with expectation at most $n(np)^{-r}$ (cf. (6)) and the corollary follows. □

3.2 The k-center problem

The k-center problem is dual to the r-dominating set problem. Given $k \in \mathbb{N}$, the aim is to find a set of k vertices which have the least maximal distance to the other vertices. For a graph G, this is just

$$r = r(G, k) = \min\{r : k \geq \gamma_r(G)\}. \tag{8}$$

From Corollary 5 it follows that whp. $r(G_{n,p}, k) = \frac{-\log pk}{\log pn} + 1 + o(1)$. The following procedure makes use of the greedy algorithm for r-independent set to compute a k-center solution:

Procedure GREEDY-k-CENTER

Input: $G, k \in \mathbb{N}$
Initialization: $r \leftarrow 0$;
Iteration: repeat
$\quad\quad\quad\quad r \leftarrow r + 1$;
$\quad\quad\quad\quad$ run GREEDY-r-INDSET on G with output W;
$\quad\quad\quad\quad$ until $|W| \leq k$;
Output: W and $r_{greedy}(G, k) = r$

We remark that the procedure RANDOMSELECT-r-DOMSET from the previous section could also be used to obtain the same results. One advantage in using GREEDY-r-INDSET is that the k-center solution that is determined is also r-independent.

Theorem 8. *Let $p = p(n)$ with $p = \omega(\log n/n)$ be given.*
(a) For any $k \in \mathbb{N}$, $k \leq n$ whp.,

$$r(G_{n,p}, k) \leq r_{greedy}(G_{n,p}, k) \leq r(G_{n,p}, k) + 1.$$

(b) There exists a set $\mathcal{A} \subseteq \{1, \ldots, n\}$ with $|\mathcal{A}| = o(n)$ such that for any $k \in \{1, \ldots, n\} \backslash \mathcal{A}$ whp.,

$$r_{greedy}(G_{n,p}, k) = r(G_{n,p}, k).$$

Proof. Let $\varepsilon > 0$. Using (8) and Corollary 5, we have that whp.

$$r(G_{n,p}, k) \geq \min\{r \in \mathbb{N} : k \geq (1 - \varepsilon/4)\frac{r \log np}{(np)^{r-1}b}\},$$

$$r_{greedy}(G_{n,p}, k) \leq \min\{r \in \mathbb{N} : k \geq (1 + \varepsilon/4)\frac{r \log np}{(np)^{r-1}b}\}.$$

As $np \geq \log n$ these inequalities imply (a).
To prove (b), let $r_0 := \max\{r : (1 - \varepsilon/4)\frac{r \log np}{(np)^{r-1}b} \leq n\}$. Let $\mathcal{A}$ be the union of intervals where r and r_{greedy} may differ, i.e.

$$\mathcal{A} = \bigcup_{r=1}^{r_0}\{(1 - \varepsilon/4)\frac{r \log np}{(np)^{r-1}b}, \ldots, (1 + \varepsilon/4)\frac{r \log np}{(np)^{r-1}b}\},$$

Then unless $k \in \mathcal{A}$, we have $r = r(G_{n,p}, k)$ with high probability. Now (b) follows from

$$|\mathcal{A}| \leq \frac{1}{2}\varepsilon \sum_{r=1}^{r_0} \frac{r \log np}{(np)^{r-1}b} \leq \frac{1}{2}\varepsilon \frac{n}{1 - \varepsilon/4}(1 + 1/(\log n) + 1/(\log n)^2 + \ldots) \leq \varepsilon n,$$

if n is large enough and $\varepsilon < 1$. □

Acknowledgement The authors are grateful to Thomas Emden-Weinert for fruitful discussions.

References

1. N. Alon, J. H. Spencer, and P. Erdős. *The Probabilistic Method.* Wiley, 1992.
2. B. Bollobás. *Random Graphs.* Academic Press, 1985.
3. Y. D. Burtin. On extreme metric parameters of a random graph: I. asymptotic estimates. *Theory of Probability and its Applications,* 19(4):710–725, 1974.
4. Y. D. Burtin. On extreme metric characteristics of a random graph: II. limit distributions. *Theory of Probability and its Applications,* 20(1):83–101, 1975.
5. G. J. Chang and G. L. Nemhauser. The k-domination and k-stability problems on sun-free chordal graphs. *SIAM Journal on Algebraic Discrete Methods,* 5(3):332–345, 1984.
6. A. Frieze and C. McDiarmid. Algorithmic theory of random graphs. *Random Structures & Algorithms,* 10:5–42, 1997.
7. A. M. Frieze. On the independence number of random graphs. *Discrete Mathematics,* 81:171–175, 1990.
8. G. R. Grimmett and C. McDiarmid. On colouring random graphs. *Proceedings of the Cambridge Philosophical Society,* 77:313–324, 1975.
9. S. T. Hedetniemi and R. C. Laskar. Bibliography on domination in graphs and some basic definitions of domination parameters. *Discrete Mathematics,* 86:257–277, 1990.
10. D. S. Hochbaum. Easy solutions for the k-center problem or the dominating set problem on random graphs. *Annals of Discrete Mathematics,* 25:189–210, 1985.
11. C. McDiarmid. Colouring random graphs. *Annals of Operations Research,* 1:183–200, 1984.

Nearly Linear Time Approximation Schemes for Euclidean TSP and Other Geometric Problems
Abstract

Sanjeev Arora[1]

Princeton University, arora@cs.princeton.edu *

Abstract. We present a randomized polynomial time approximation scheme for Euclidean TSP in $\Re^2$ that is substantially more efficient than our earlier scheme in *Polynomial-time Approximation Schemes for Euclidean TSP and other Geometric Problems*, Proceedings of IEEE FOCS 1996, pp 2-11 (and the scheme of Mitchell, which he discovered a few months later). For any fixed $c > 1$ and any set of n nodes in the plane, the new scheme finds a $(1 + \frac{1}{c})$-approximation to the optimum traveling salesman tour in $O(n(\log n)^{O(c)})$ time. (Our earlier scheme ran in $n^{O(c)}$ time.) For points in $\Re^d$ the algorithm runs in $O(n(\log n)^{(O(\sqrt{d}c))^{d-1}})$ time. This time is polynomial (actually nearly linear) for every fixed c, d; designing such a polynomial-time algorithm was an open problem (our earlier algorithm ran in quasipolynomial time for $d \geq 3$).

The algorithm generalizes to the same set of Euclidean problems handled by the previous algorithm, including Steiner Tree, k-TSP, Minimum degree-restricted spanning tree, k-MST, etc, although for k-TSP and k-MST the running time gets multiplied by k.

We also use our ideas to design nearly-linear time approximation schemes for Euclidean versions of problems that are known to be in P, such as Minimum Spanning Tree and Min Cost Perfect Matching.

All our algorithms also work, with almost no modification, when distance is measured using any geometric norm (such as ℓ_p for $p \geq 1$ or other Minkowski norms). They also have simple parallel implementations (say, in NC^2).

We design the PTAS by showing that the plane can be recursively partitioned (using a randomized variant of the *quadtree*) such that some $(1 + 1/c)$-approximate salesman tour crosses each line of the partition at most $r = O(c)$ times Such a tour can be found by dynamic programming. For each line in the partition the algorithm first "guesses" where the tour crosses this line and the order in which those crossings occur. Then the algorithm recurses independently on the two sides of the line. There are only $n \log n$ distinct regions in the partition. Within each region it is sufficient to "guess" the places where the tour crosses the boundary quite coarsely. The dynamic programming then takes only $(O(\log n))^{O(r)} = (\log n)^{O(c)}$ time per region, for a total running time of $n \cdot (\log n)^{O(c)}$.

* Supported by NSF CAREER award NSF CCR-9502747 and an Alfred Sloan Fellowship

Random Sampling of Euler Tours

Prasad Tetali[1] and Santosh Vempala[2]

[1] School of Mathematics, Georgia Institute of Technology, Atlanta GA 30332,
(research supported by the NSF Grant No. CCR-9503952) tetali@math.gatech.edu
[2] School of Computer Science, Carnegie Mellon University, Pittsburgh PA 15213
svempala@cs.cmu.edu

Abstract. We define a Markov chain on the set of Euler tours of a given Eulerian graph based on transformations first defined by Kotzig in 1966. We prove that the chain is rapidly mixing if the maximum degree in the given graph is 6, thus obtaining an efficient algorithm for sampling and counting the set of Euler tours for such an Eulerian graph.

1 Introduction

Let $G = (V, E)$ denote an undirected Eulerian graph; i.e. the degree $d(v)$ of each vertex $v \in V$ is even. We allow G to have multiple edges, but not self-loops. We consider the problem of generating an Euler tour uniformly at random from among the set $\mathrm{EUL}(G)$ of all Euler tours of G. Let $\mathrm{eul}(G)$ denote the cardinality of $\mathrm{EUL}(G)$. Note that there is a simple linear time algorithm for constructing one such tour (see e.g. any basic book on design and analysis of algorithms). However, there is no known efficient algorithm to count the *exact* number, $\mathrm{eul}(G)$, of them nor is it known that computing $\mathrm{eul}(G)$ is #P-hard.

In [4], Kotzig introduced a simple "local" transformation and showed that any Euler tour of a 4-regular graph can be transformed into any other Euler tour by a finite sequence of such transformations. In [1], this result was extended to all Eulerian graphs. In this work we describe a simple Markov chain with state space $\mathrm{EUL}(G)$ and transitions defined by Kotzig-transformations (κ-transformation). The theorem of [1] shows that this chain is irreducible (i.e. the graph underlying the Markov chain is connected). We investigate the *mixing* properties of the Markov chain. We are able to prove that this chain is rapidly mixing as long as $d(v) \leq 6$, for all $v \in V$. Our proof is based on *canonical path counting*, a technique first used by Jerrum and Sinclair [8].

Counting the number of *directed* Euler tours in an Eulerian *digraph* is a classical result (see e.g. [10]), based on a certain determinantal evaluation. However, the undirected case has remained an intractable problem for a number of years. (Thus this problem has the interesting feature that it is seemingly easier for directed graphs than for the undirected.) We hope that our result provides a breakthrough, and revives interest in this classical topic. McKay and Robinson in a recent paper [6] obtained an asymptotic enumeration of the number of tours in the complete graph (K_n, for n odd). Also implicit in their results is an efficient probabilistic algorithm which ouputs an Euler tour of K_n uniformly at random with high probability.

The only other related results which we are aware of are the following. An Eulerian orientation of an undirected Eulerian graph is an orientation of the edges such that every vertex has indegree equalling the outdegree. In [7], Mihail and Winkler showed that counting the exact number of them in general is #P-complete, and also provided an *fpras* for an approximate count using the Markov chain Monte-Carlo method.

2   Kotzig transformations and the Markov chain on Euler tours

Before we state and prove our results more precisely, a few clarifications are due. We do not consider cyclic rotations of the sequence of edges in a particular tour T as different from T. Also an Euler tour and its complete reversal (of the sequence of its edges) are treated as the same tour.

An Euler tour T of G passes through a vertex v of degree $2k$ exactly k times, and T can be thought of as an alternating sequence of vertices and edges. We refer to any part of T between two specified transitions through a given vertex v as a *segment* of T corresponding to v. (Note that a segment corresponding to v can pass through v several times.) A *Kotzig-transformation* (κ-transformation , for short) at vertex v consists of simply reversing a segment corresponding to v. That is, reversing the sequence of edges and vertices between two transitions (not necessarily consecutive) of a tour T through v. For a proof of the following elegant theorem we refer to [1]. We remark that the proof is by induction on the maximum degree of G. The base case of the induction is the 4-regular case which was first proved in [4]; a slightly different proof of the 4-regular case is described in Section 3.1 below.

Theorem 1 (Abrham-Kotzig). *Let T and T' be two Euler tours of an Eulerian multigraph G. Then there exists a finite sequence of κ-transformations at the vertices of G which transforms T into T'.*

It is now natural to consider the following Markov chain with the set of Euler tours as the state space and the transitions corresponding to κ-transformations ; i.e. pick a vertex u.a.r. and apply a (random one, if there is more than one) κ-transformation at v with probability $1/2$, and with the remaining probability *do nothing*. Since we do not have a proof that this chain is rapidly mixing in general, for the most part of the rest of the paper we shall restrict ourselves to the particularly elegant case of 4-regular Eulerian graphs; in Section 4, we describe the extension to yield rapid-mixing for graphs with maximum degree 6.

2.1 A detour : κ-transformation and transition systems

It turns out to be convenient to have the following representation of an Euler tour. This was suggested in [4] and [1]. Let $V = \{v_1, \ldots, v_n\}$ be the set of vertices of G. Then a *transition system* T of G is defined as $T = \{T(v_1), \ldots, T(v_n)\}$, where $T(v_i)$ is an arbitrary pairing of the edges incident at v_i. By a *pairing* of edges here, we mean partitioning of edges incident at a vertex into (unordered) pairs of edges. It is easy to see that each Euler tour of G induces a unique transition system, but not every transition system corresponds to an Euler tour. (Otherwise, counting the number of Euler tours would have been an easy task!) A pair of edges $\{e, e'\}$ at a vertex v in such a pairing (also called a *transition*), tells us that if a tour with this pairing enters v via e, then it must exit via e' and vice versa. Also note that a κ-transformation at a vertex v corresponds simply to an interchange of two edges from two pairs of $T(v)$, where T is the transition system corresponding to the Euler tour at hand. In view of this correspondence, we may use T to refer to both an Euler tour and its transition system.

Given two transition systems T, T', we use $T \Rightarrow_v T'$ to denote that we can transform $T(v)$ into $T'(v)$ by a κ-transformation at v and call this an *allowed transition*; to denote the contrary, we use $T \not\Rightarrow_v T'$, and call it a *prohibited transition*. The latter might happen, for example, when T is an Euler tour and T' is a transition system but not an Euler tour.

3 The 4-regular case

Let G be a 4-regular graph, where multiple edges are allowed. In this case the pairings of a transition system induced from an Euler tour take a simple form – each vertex has three possible pairings of its edges. For each pairing, precisely one other pairing corresponds to an allowed transition (i.e. the result of making a κ-transformation), and the other pairing corresponds to a prohibited transition. We make the following easy but crucial facts about the effect of κ-transformations on pairings. Let A be an Euler tour of G, and consider a κ-transformation on A.

Fact 1: a κ-transformation at u does not change $A(v)$, for $v \neq u$.
Fact 2: a κ-transformation at u, however, might swap the roles of the allowed and the prohibited transitions of $A(v)$. This happens precisely when the visits to u and v are *interleaved* (or "alternating") in tour A.
Note that the Markov chain for the 4-regular case takes the following simple form:

- Start with an arbitrary Euler tour of G.

- At each step, pick a vertex v u.a.r. and with probability $1/2$ apply the unique possible (with respect to the current tour) κ-transformation at v, and with the remaining probability do nothing.

3.1 Canonical Paths

We now proceed to describe a canonical path between every pair of states in our Markov chain. Recall that the steps of our Markov chain are κ-transformations as described previously.

Let A and B denote two distinct Euler tours of G. We first describe in the following a finite sequence of κ-transformations which transforms $A = A_0$ to $A_1, A_2, \ldots, A_m$, finally to B, such that $A_j(v_i) = B(v_i)$, for all $i \leq j - 1$, and for $j = 1, \ldots, m$; as we shall see below $v_1, \ldots, v_n$ is going to be an ordering of the vertices which depends on the *symmetric difference* of A and B. The basic idea is in the following lemma.

Lemma 2. *Let A and B be two Euler tours of a 4-regular graph G. Suppose $A(v) \neq B(v)$, and $A \not\Rightarrow_v B$, then*

1. *there exists $u(\neq v)$ such that $A(u) \neq B(u)$, and*

2. *in at most 3 κ-transformations (on A) we can transform A to A' so that $A'(u) = B(u)$, $A'(v) = B(v)$, and $A'(w) = A(w)$, for $w \neq u, v$.*

Proof. The proof is illustrated (see Fig. 1) without loss of generality by a 2-vertex multigraph with the edges e, e', f and f'. Let (e, e') and (f, f') be the two transitions of a tour A at some vertex v. Assume we can reach the tour which has (e, f) and (e', f') at v via a k-transformation. This implies that A has segments $S_1 = e \ldots f'$ and $S_2 = e' \ldots f$ at v. Let B be a tour with the pairs $(e, f'), (e', f)$ at v. Thus $A \not\Rightarrow_v B$. Then there must be some vertex u such that the segments S_1, S_2 of A meet at u, and such that the pairing at u in B is different from that in A (i.e. $A(u) \neq B(u)$). This is because if A and B only differ at v in a prohibited way, then one of them *must not* be an Euler tour.

To see the second part of the conclusion of the lemma, suppose $A \Rightarrow_u B$. Now, by Fact 2 above, the pairings $(e, f'), (e'f)$ are no longer prohibited for $A(v)$ (the segments at v are different). So we can make another κ-transformation to make the pairing at v as in B. If the suggested κ-transformation at u is prohibited (i.e. $A \not\Rightarrow_u B$) then first we apply a κ-transformation at v that gives the pairing $(e, f), (e', f')$ at v (this is not prohibited in A). Now we can apply the required κ-transformation at u, followed by another at v to match up with B at both u and v. □

For each $A \triangle B$, we define an ordering on the vertices of G, thus completely specifying the canonical path from A to B. The vertices which do not appear in $A \triangle B$ (at which no κ-transformation is required) can be ordered arbitrarily, and we fix them to be the smallest elements in the ordering. The vertices which appear in $A \triangle B$ are ordered one or two (as consecutive elements) at a time according to the following rules.

Rule 1. Suppose w is such that $A \Rightarrow_w B$ but $B \not\Rightarrow_w A$. (This does not concern us as far as going from tour A to tour B is concerned; however, this is pertinent to the *complementary path* to be used below, wherein we need to be able to go from B to A.) In this case, we know (from the proof of Lemma 2) that there exists another vertex w' which appears in $A \triangle B$ such that $B \Rightarrow_{w'} A$ and making

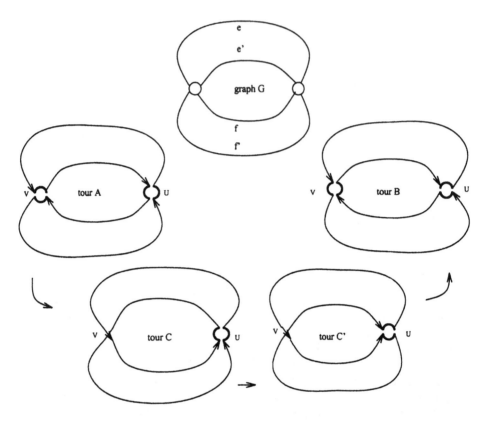

Fig. 1. Kotzig transformations transforming $A \Rightarrow_V C \Rightarrow_U C' \Rightarrow_V B$

that κ-transformation at w' in B allows $B \Rightarrow_w A$. Thus with foresight we order the pair w, w' such that w immediately precedes w'.

Rule 2. Now we consider u such that $A \not\Rightarrow_u B$. By Lemma 2, this involves a pair u, v such that using two or three κ-transformations we can change $A(u)$ to $B(u)$ and $A(v)$ to $B(v)$. If we can make A and B agree at both v and u in two κ-transformations then we let v precede u, otherwise (meaning we have to make three κ-transformations , first at v, the second at u, and finally the third at v) u precedes v.

Rule 3. If w is such that $A \Rightarrow_w B$ and $B \Rightarrow_w A$, then w appears by itself in the ordering.

We repeat this till all the vertices of G have been ordered. Such an ordering is chosen apriori for every pair A, B of Euler tours. Thus the canonical path from A to B simply involves applying κ-transformations on the vertices of G using such an ordering defined by $A \triangle B$.

We treat the steps described in the above lemma as a *phase*. So in one phase we move to a tour A' such that the $A'\Delta B$ is a proper subset of $A\Delta B$ and moreover the disjoint union of $A\Delta A'$ and $A'\Delta B$ is $A\Delta B$. It is important to note that an intermediate tour C on such a path can have at most one vertex v where the pairing $C(v)$ is different from both $A(v)$ and $B(v)$. This happens in the second case of a phase as described above.

We get the following result of [2] as an immediate consequence of the Lemma 2. The proof in [2] is not very transparent. Fon der Flaass [3] improved the bound in the following result to $10n/9$.

Corollary 3 (Bouchet). *Let A and B be two Euler tours of a 4-regular graph. Then A can be transformed into B by applying a sequence of at most $3n/2$ κ-transformations .*

3.2 Proof of rapid mixing

For Euler tours A, B, let $A\Delta B$ denote the symmetric difference of their transition systems. Thus $A\Delta B = \{A\Delta B(v) : v \in V\}$, where $A\Delta B(v) = A(v)\backslash B(v)\cup B(v)\backslash A(v)$. Let γ_{AB} denote the canonical path from A to B, and let $\Gamma = \{\gamma_{A,B} : A, B \in M(G)\}$ denote the set of all canonical paths. For an arbitrary transition (C, C') of the Markov chain, let $\Gamma(C, C') = \{(A, B) : \gamma_{A,B} \text{ uses the transition } (C, C')\}$ denote the set of (ordered) pairs (A, B) of tours such that the canonical path from A to B goes through (C, C').

The complementary tour. To bound the number of paths through a step (C, C') of our Markov chain, we define a *complementary* tour D so as to achieve the following. In general, at a vertex where the pairing in C agrees with that in A we let D have the pairing in B, and vice versa, and at vertices where A and B agree (and so does C), we have in D the same pairing. At the single vertex where C might differ from both A and B (in view of Rule 2 above) we let D agree with C. If C is obtained by a κ-transformation as specified by Rule 1 above, then we will have C and D agree at the two vertices (w, w') that are ordered consecutively. *Thus there might be at most two vertices at which A and B differ but C and D agree.* All of this can be accomplished in a natural way as follows.

Let $k_1, k_2, \ldots, k_m$ be the sequence of κ-transformations on the canonical path from A to B. For the purpose of defining the tour D, we define a path from B to A which is *complementary* to our canonical path from A to B. There are three cases to be considered depending on the three rules introduced in defining the canonical paths. First let k_1 and k_2 be the κ-transformations applied on A at vertices w and w' as in Rule 1. Then define k_1^{-1} and k_2^{-1} to be the κ-transformations applied on B at w' and w respectively. (Note that this makes the inverses valid transformations.) Next, if k_1 changes the pairing in A to one that is different from both A and B (as in Rule 2) then let k_1^{-1} be the κ-transformation that changes the pairing in B at v to the pairing that is different from both A and B. Finally assume k_1 changes the pairing in A at some vertex v to match up with the pairing in B as in Rule 3. In this case k_1^{-1} is the κ-transformation on B that changes the pairing at v to agree with the pairing in

A. In all cases the inverses are valid κ-transformations , since we defined the canonical paths with this foresight!

Consider an intermediate tour C that is reached from A via the sequence $k_1, k_2, \ldots, k_m$. Construct D from B using the sequence $k_1^{-1}, k_2^{-1}, \ldots, k_m^{-1}$. Note that D is an Euler tour since it is reachable (from B) via κ-transformations . Thus by setting $\sigma_{(C,C')}(A, B) := D$, we have a genuine map $\sigma_{(C,C')} : \Gamma(C, C') \to$ EUL(G).

Lemma 4. *Let* (C, C'), (A, B), *and* D *be as defined above. Given* (C, C'), *there is a mapping from* D *to pairs* (A, B) *that maps a single* D *to at most* $n(n-1)$ *pairs.*

Proof. First note that there are at most $n(n-1)$ choices for $A\Delta B$, given C, C', and D. For, $A\Delta B = C\Delta D$, everywhere except for perhaps at two vertices – a vertex w where $B\not\approx_w A$ and a vertex v where $A\not\approx_v B$. At both vertices we have that A and B disagree and C and D agree with each other. In any case, there are at most $n(n-1)$ choices for the two vertices. And once we fix a choice, $A\Delta B$ is determined at that choice and hence at all vertices. (Note that this uses very much the fact that the graph is 4-regular.)

Now assuming we know $A\Delta B$, besides knowing the current transition (C, C') and D (as defined above), we show how A and B can be reconstructed uniquely. Since we know $A\Delta B$, we also know the (canonical) ordering defined by $A\Delta B$. Let the ordering be $(u_1, \ldots, u_k, \ldots, u_n)$. If u_k is the current vertex (i.e. involved in (C, C')), then the required changes (from the pairing in A to the pairing in B) have already been made at $u_1, \ldots, u_{k-2}$, and no changes have been made as yet at $u_{k+2}, \ldots, u_n$. Thus we know $A(v)$ and $B(v)$ at all v, except for $v \in \{u_{k-1}, u_k, u_{k+1}\}$.

Case 1. If $C(u_k) \neq D(u_k)$, then set $A(u_k) := C(u_k)$ and $B(u_k) := D(u_k)$.

In addition, if $C'(u_k) \notin A\Delta B(u_k)$, then set $A(u_{k-1}) := C(u_{k-1})$, and $B(u_{k-1}) := D(u_{k-1})$, since this only arises in the situation when we apply a κ-transformation at u_k before applying a κ-transformation at u_{k-1} (as in the second case of a phase). We also set $A(u_{k+1}) := C(u_{k+1})$ and $B(u_{k+1}) := D(u_{k+1})$.

If $C'(u_k) \in A\Delta B(u_k)$, this implies a valid κ-transformation at u_k, having followed either (a) a valid κ-transformation at u_{k-1}, or (b) a forbidden κ-transformation at u_{k+1}. In either case, we set $A(u_{k-1}) := D(u_{k-1})$, and $B(u_{k-1}) := C(u_{k-1})$. Note that at this stage, we only need to determine A and B at u_{k+1}. If $C(u_{k+1}) \in A\Delta B(u_{k+1}) \neq \phi$, then it must have been case (a) and we set $A(u_{k+1}) := C(u_{k+1})$ and $B(u_{k+1}) := D(u_{k+1})$. Otherwise, it must have been case (b) and we can set $A(u_{k+1})$ (and $B(u_{k+1})$) to the unique pairing different from $C(u_{k+1})$ which would make the transition system of A (and B) an Euler tour. (Note that once again we have made heavy use of the fact that the graph is 4-regular.)

Case 2. If $C(u_k) = D(u_k)$, then u_k is such that either (a) $B\not\approx_{u_k} A$ or (b) $A\not\approx_{u_k} B$. In either case, we can set $B(u_k) := C'(u_k)$, $A(u_{k+1}) := C(u_{k+1})$, and $B(u_{k+1}) := D(u_{k+1})$. Moreover, if $C(u_{k-1}) = D(u_{k-1})$ then it must have been case (a), otherwise it must have been case (b).

If case (a) then we set $A(u_k) := C(u_k)$ and $B(u_{k-1}) := C(u_{k-1})$, and now that B is completely determined, knowing $A \triangle B$, we can determine A. In particular $A(u_{k-1})$ can be determined.

If case (b) then we set $A(u_{k-1}) := D(u_{k-1})$ and $B(u_{k-1}) := C(u_{k-1})$. Since B is now completely determined we can determine A; in particular, in this case $A(u_k)$ gets the unique pairing different from both $B(u_k)$ and $C(u_k)$. □

As a consequence of the above lemma we have a bound on the conductance.

Theorem 5. *The conductance of the Markov chain on Euler tours is at least* $\frac{1}{2n^2}$.

4 The case of maximum degree six

Given two Euler tours A, B of a graph in which the degree of a vertex at most 6, the pairing at a vertex v in the two tours could agree completely (i.e. $A(v) = B(v)$), or agree in 1 pair, which we denote as $A(v) =_1 B(v)$, or differ in all 3 pairs (i.e. $A(v) \neq B(v)$).

Consider now the κ-transformations at v on A.

Lemma 6. *Let A and B be two Euler tours of a graph G with degree at most 6. Then,*

1. *suppose $A(v) =_1 B(v)$, and $A \not\to_v B$ then there exists $u(\neq v)$ such that $A(u) \neq B(u)$ or $A(u) =_1 B(u)$, and in at most 3 κ-transformations (on A) we can transform A to A' so that $A'(u) =_1 B(u)$ or $A'(u) = B(u)$, and $A'(v) = B(v)$, and $A'(w) = A(w)$, for all $w \neq u, v$.*

2. *suppose $A(v) \neq B(v)$, and any κ-transformation at v on A that would result in $A(v) =_1 B(v)$ is prohibited, then there exists a vertex u such that $A(u) \neq B(u)$ or $A(u) =_1 B(u)$, and in at most 3 κ-transformations (on A) we can transform A to A' so that $A'(u) =_1 B(u)$ or $A'(u) = B(u)$, and $A'(v) =_1 B(v)$, and $A'(w) = A(w)$, for all $w \neq u, v$.*

Proof. Similar to Lemma 2 □

So to transform $A(v)$ to $B(v)$ might involve up to 6 κ-transformations and result in up to *two* vertices u, w such that $A'(u) =_1 B(u)$ and $A'(w) =_1 B'(w)$. Thus an intermediate tour C (A' in this case) might have two vertices where its pairing differs from that of both A and B. The crucial observation that allows us to prove rapid-mixing for the degree 6 case is that this is actually the maximum number. In the canonical path we now focus our attention on one of u or w, say u. Then by the first case of the above lemma, we can move to a tour A'' which agrees with B at u, and in the process we create at most one other vertex where $A''(w) =_1 B(w)$, keeping the number of vertices where we have partially "fixed" the tour at no more than 2.

To completely specify the path from A to B we choose a suitable ordering on the vertices of G, placing the vertices not in $A \triangle B$ at the beginning.

Lemma 7. *An intermediate tour C on the canonical path from A to B can have at most two vertices u, w where its pairing differs from both A and B.*

The proof of low congestion, (and hence high conductance) can be carried out in a fashion almost identical to the proof of Lemma 4 in the previous section. The bound we get here is weaker.

Theorem 8. *The conductance of the Markov chain for a graph with degree at most 6 is $\Omega(\frac{1}{n^2})$.*

5 Conclusion

Is this Markov chain rapidly mixing in general? In particular, can we extend this approach to graphs with higher degrees? One difficulty in trying to do so is that the number of vertices in intermediate tours that are different from both the source and the destination seems to be unbounded for 8-regular and higher degree graphs. It is worth mentioning that an alternate proof based on coupling might provide some insight for the general case.

References

1. J. Abrham and A. Kotzig, Transformations of Euler tours, *Annals of Discrete Mathematics* **8** (1980), 65-69.

2. A. Bouchet, κ-transformations , local complementations, and Switchings, in Hahn, G., Sabidussi, G., Woodrow, R. (eds.), "Cycles and Rays," NATO ASI Ser. C, Kluwer Academic Publ., Dordrecht (1990), 41-50.

3. D. Fon der Flaass, Distance between locally equivalent graphs, *Metody Diskretnogo Analiza*, Novosibirsk **48**(1989), 85-94. [In Russian].

4. A. Kotzig, Eulerian lines in finite 4-valent graphs and their transformations, in: P.Erdős and G. Katona, Eds., "Theory of Graphs," Proc. of the Colloq. held at Tihany, Hungary (1966) (Akademiai Kiado, Publishing House of the Hungarian Academy of Sciences, Budapest (1968), 219-230).

5. B.D. McKay, The asymptotic number of regular tournaments, eulerian digraphs and eulerian oriented graphs, *Combinatorica*, **10** (1990), 367-377.

6. B.D. McKay and R. Robinson, Asymptotic enumeration of Eulerian circuits in the complete graph, preprint (1995).

7. M. Mihail and P. Winkler, On the number of Eulerian orientations of a graph, *Proc. of the 3rd ACM-SIAM Symp. on Discrete Algorithms* (1992), 138-145.

8. M.R. Jerrum and A.J. Sinclair, The Markov chain Monte Carlo method, in "Approximation algorithms for NP-hard problems," D.S. Hochbaum (ed.), PWS Publishing, Boston, 1997.

9. A.J. Sinclair, "Algorithms for random generation & counting : a Markov chain approach," Progress in Theoretical Computer Science, Birkhäuser (1992).

10. L. Lovász, "Combinatorial Problems and Exercises," 1993 (second edition), North-Holland, Elsevier Science Publishers (Amsterdam) and Akade'miai Kiado' (Budapest).

A Combinatorial Consistency Lemma
with Application to Proving the PCP Theorem

Oded Goldreich[1] and Shmuel Safra[2]

[1] Department of Computer Science and Applied Mathematics, Weizmann Institute of Science, Rehovot, ISRAEL. E-mail: oded@wisdom.weizmann.ac.il.
[2] Computer Science Department, Sackler Faculty of Exact Sciences, Tel-Aviv University, Ramat-Aviv, ISRAEL. Email: safra@math.tau.ac.il.

Abstract. The current proof of the PCP Theorem (i.e., $\mathcal{NP} = \mathcal{PCP}(\log, O(1))$) is very complicated. One source of difficulty is the technically involved analysis of low-degree tests. Here, we refer to the difficulty of obtaining *strong* results regarding low-degree tests; namely, results of the type obtained and used by Arora and Safra and Arora et. al.

In this paper, we eliminate the need to obtain such strong results on low-degree tests when proving the PCP Theorem. Although we do not get rid of low-degree tests altogether, using our results it is now possible to prove the PCP Theorem using a simpler analysis of low-degree tests (which yields weaker bounds). In other words, we replace the strong algebraic analysis of low-degree tests presented by Arora and Safra and Arora et. al. by a combinatorial lemma (which does not refer to low-degree tests or polynomials).

Keywords: Parallelization of Probabilistic Proof Systems, Probabilistically Checkable Proofs (PCP), NP, Low-Degree Tests.

1   Introduction

The characterization of $\mathcal{NP}$ in terms of Probabilistically Checkable Proofs (PCP systems) [AS, ALMSS], hereafter referred to as the **PCP Characterization Theorem**, is one of the more fundamental achievements of complexity theory. Loosely speaking, this theorem states that membership in any NP-language can be verified probabilistically by a polynomial-time machine which inspects a constant number of bits (in random locations) in a "redundant" NP-witness. Unfortunately, the current proof of the PCP Characterization Theorem is very complicated and, consequently, it has not been assimilated into complexity theory. Clearly, changing this state of affairs is highly desirable.

There are two things which make the current proof (of the PCP Characterization Theorem) difficult. One source of difficulty is the complicated conceptual structure of the proof (most notably the acclaimed 'proof composition' paradigm). Yet, with time, this part seems easier to understand and explain than when it was first introduced. Furthermore, the Proof Composition Paradigm turned out to be very useful and played a central role in subsequent works in this area (cf., [BGLR, BS, BGS, H96]). The other source of difficulty is the technically involved analysis of low-degree tests. Here we refer to the difficulty of obtaining *strong* results regarding low-degree tests; namely, results of the type obtained and used in [AS] and [ALMSS].

In this paper, we eliminate the latter difficulty. Although we do not get rid of low-degree tests altogether, using our results it is now possible to prove the PCP Characterization Theorem using only the weaker and simpler analysis of low-degree tests presented in [GLRSW, RS92, RS96]. In other words, we replace the complicated algebraic analysis of low-degree tests presented in [AS, ALMSS] by a combinatorial lemma (which does not refer to low-degree tests or even to polynomials). We believe that this combinatorial lemma is very intuitive and find its proof much simpler than the algebraic analysis of [AS, ALMSS]. (However, simplicity may be a matter of taste.)

Loosely speaking, our combinatorial lemma provides a way of generating sequences of pairwise independent random points so that any assignment of values to the sequences must induce consistent values on the individual elements. This is obtained by a "consistency test" which samples a constant number of sequences. We stress that the length of the sequences as well as the domain from which the elements are chosen are parameters, which may grow while the number of samples remains fixed.

1.1 Two Combinatorial Consistency Lemmas

The following problem arises frequently when trying to design PCP systems, and in particular when proving the PCP Characterization Theorem. For some sets S and V, one has a procedure, which given (bounded) oracle access to any function $f : S \mapsto V$, tests if f has some desired property. Furthermore, in case f is sufficiently bad (i.e., far from any function having the property), the test detects this with "noticeable" probability. For example, the function f may be the proof-oracle in a basic PCP system which we want to utilize (as an ingredient in the composition of PCP systems). The problem is that we want to increase the detection probability (equivalently, reduce the error probability) without increasing the number of queries, although we are willing to allow more informative queries. For example, we are willing to allow queries in which one supplies a sequence of elements in S and expects to obtain the corresponding sequence of values of f on these elements. The problem is that the sequences of values obtained may not be consistent with any function $f : S \mapsto V$.

We can now phrase a simple problem of testing consistency. One is given access to a function $F : S^\ell \mapsto V^\ell$ and is asked whether there exists a function $f : S \mapsto V$ so that for most sequences $(x_1, ..., x_\ell) \in S^\ell$,

$$F(x_1, ..., x_\ell) = (f(x_1), ..., f(x_\ell)) \, .$$

Loosely speaking, we prove that querying F on a constant number of related random sequences suffices for testing a relaxion of the above. That is,

Lemma 1.1 (combinatorial consistency – simple case): *For every $\delta > 0$, there exists a constant $c = \text{poly}(1/\delta)$ and a probabilistic oracle machine, T, which on input $(\ell, |S|)$ runs for $\text{poly}(\ell \cdot \log |S|)$-time and makes at most c queries to an oracle $F : S^\ell \mapsto V^\ell$, such that*

- *If there exist a function $f : S \mapsto V$ such that $F(x_1, ..., x_\ell) = (f(x_1), ..., f(x_\ell))$, for all $(x_1, ..., x_\ell) \in S^\ell$, then T always accepts when given access to oracle F.*
- *If T accepts with probability at least $\frac{1}{2}$, when given access to oracle F, then there exist a function $f : S \mapsto V$ such that the sequences $F(x_1, ..., x_\ell)$ and $(f(x_1), ..., f(x_\ell))$ agree on at least $\ell - \sqrt{\ell}$ positions, for at least a $1 - \delta$ fraction of all possible $(x_1, ..., x_\ell) \in S^\ell$.*

Specifically, the test examines the value of the function F on random pairs of sequences $((r_1, ..., r_\ell), (s_1, ..., s_\ell))$, where $r_i = s_i$ for $\sqrt{\ell}$ of the i's, and checks that the corresponding values (on these r_i's and s_i's) are indeed equal. For details see Section 4.

Unfortunately, this relatively simple consistency lemma does not suffice for the PCP applications. The reason being that, in that application, error reduction (see above) is done via randomness-efficient procedures such as pairwise-independent sequences (since we cannot afford to utilize $\ell \cdot \log_2 |S|$ random bits as above). Consequently, the function F is not defined on the entire set S^ℓ but rather on a very sparse subset, denoted $\mathbf{S}$. Thus, one is given access to a function $F : \mathbf{S} \mapsto V^\ell$ and is asked whether there exists a function $f : S \mapsto V$ so that for most sequences $(x_1, ..., x_\ell) \in \mathbf{S}$, the sequences $F(x_1, ..., x_\ell)$ and $(f(x_1), ..., f(x_\ell))$ agree on most (continuous) subsequences of length $\sqrt{\ell}$. The main result of this paper is

Lemma 1.2 (combinatorial consistency – sparse case): *For every two of integers $s, \ell > 1$, there exists a set $\mathbf{S}_{s,\ell} \subset [s]^\ell$, where $[s] \overset{\text{def}}{=} \{1, ..., s\}$, so that the following holds:*

1. *For every $\delta > 0$, there exists a constant $c = \text{poly}(1/\delta)$ and a probabilistic oracle machine, T, which on input (ℓ, s) runs for $\text{poly}(\ell \cdot \log s)$-time and makes at most c queries to an oracle $F : \mathbf{S}_{s,\ell} \mapsto V^\ell$, such that*
   - *If there exist a function $f : [s] \mapsto V$ such that $F(x_1, ..., x_\ell) = (f(x_1), ..., f(x_\ell))$, for all $(x_1, ..., x_\ell) \in \mathbf{S}_{s,\ell}$, then T always accepts when given access to oracle F.*
   - *If T accepts with probability at least $\frac{1}{2}$, when given access to oracle F, then there exist a function $f : [s] \mapsto V$ such that for at least a $1 - \delta$ fraction of all possible $(x_1, ..., x_\ell) \in \mathbf{S}_{s,\ell}$ the sequences $F(x_1, ..., x_\ell)$ and $(f(x_1), ..., f(x_\ell))$ agree on at least a $1 - \delta$ fraction of the subsequences of length $\sqrt{\ell}$.*
2. *The individual elements in a uniformly selected sequence in $\mathbf{S}_{s,\ell}$ are uniformly distributed in $[s]$ and are pairwise independent. Furthermore, the set $\mathbf{S}_{s,\ell}$ has cardinality $\text{poly}(s)$ and can be constructed in time $\text{poly}(s, \ell)$.*

Specifically, the test examines the value of the function F on related random pairs of sequences $((r_1, ..., r_\ell), (s_1, ..., s_\ell)) \in \mathbf{S}_{s,\ell}$. These sequences are viewed as $\sqrt{\ell} \times \sqrt{\ell}$ matrices, and, loosely speaking, they are chosen to be random extensions of the same random row (or column). For details see Section 2. In particular, the presentation in Section 2 axiomatizes properties of the set of sequences, $\mathbf{S}_{s,\ell}$, for which the above tester works. Thus, we provide a "parallel repetition theorem" which holds for random but non-independent instances (rather than for independent random instances as in other such results). However, our "parallel repetition theorem" applies only to the case where a single query is asked in the basic system (rather than a pair of related queries as in other results). Due to this limitation, we could not apply our "parallel repetition theorem" directly to the error-reduction of generic proof systems. Instead, as explained below, we applied our "parallel repetition theorem" to derive a relatively strong low-degree test from a weaker low-degree test. We believe that the combinatorial consistency lemma of Section 2 may play a role in subsequent developments in the area.

1.2 Application to the PCP Characterization Theorem

The currently known proof of the PCP Characterization Theorem [ALMSS] composes proof systems in which the verifier makes a constant number of multi-valued queries. Such verifiers are constructed by "parallelization" of simpler verifiers, and thus the problem of "consistency" arises. This problem is solved by use of low-degree multi-variant polynomials, which in turn requires "high-quality" low-degree testers. Specifically, given a function $f : \mathrm{GF}(p)^n \mapsto \mathrm{GF}(p)$, where p is prime, one needs to test if f is close to some low-degree polynomial (in n variables over the finite field $\mathrm{GF}(p)$). It is required that any function f which disagrees with every d-degree polynomial on at least (say) 1% of the inputs be rejected with (say) probability 99%. The test is allowed to use auxiliary proof oracles (in addition to f) but it may only make a *constant* number of queries and the answers must have length bounded by $\mathrm{poly}(n, d, \log p)$. Using a technical lemma due to Arora and Safra [AS], Arora et. al. [ALMSS] proved such a result.[3] The full proof is quite complex and is algebraic in nature. A weaker result due to Gemmel et. al. [GLRSW] (see [RS96]) asserts the existence of a d-degree test which, using $d + 2$ queries, rejects such bad functions with probability at least $\Omega(1/d^2)$. Their proof is much simpler. Combining the result of Gemmel et. al. [GLRSW, RS96] with our combinatorial consistency lemma (i.e., Lemma 1.2), we obtain an alternative proof of the following result

Lemma 1.3 (low-degree tester): *For every $\delta > 0$, there exists a constant c and a probabilistic oracle machine, T, which on input n, p, d runs for $\mathrm{poly}(n, d, \log p)$-time and makes at most c queries to both f and to an auxiliary oracle F, such that*

- *If f is a degree-d polynomial, then there exist a function F so that T always accepts.*

---

[3] An improved analysis was later obtained by Friedl and Sudan [FS].

- *If T accepts with probability at least $\frac{1}{2}$, when given access to the oracles f and F, then f agrees with some degree-d polynomial on at least a $1 - \Omega(1/d^2)$ fraction of the domain.*[4]

We stress that in contrast to [ALMSS] our proof of the above lemma is mainly combinatorial. Our only reference to algebra is in relying on the result of Gemmel et. al. [GLRSW, RS96] (which is weaker and has a simpler proof than that of [ALMSS]). Our tester works by performing many (pairwise independent) instances of the [GLRSW] test in parallel, and by guaranteeing the consistency of the answers obtained in these tests via our combinatorial consistency test (i.e., of Lemma 1.2). In contrast, prior to our work, the only way to guarantee the consistency of these answers resulted in the need to perform a low-degree test of the type asserted in Lemma 1.3 (and using [ALMSS], which was the only alternative known, this meant losing the advantage of utilizing a low-degree tests with a simpler algebraic analysis).

1.3 Related work

We refrain from an attempt to provide an account of the developments which have culminated in the PCP Characterization Theorem. Works which should certainly be mentioned include [GMR, BGKW, FRS, LFKN, S90, BFL, BFLS, FGLSS, AS, ALMSS] as well as [BF, BLR, LS, RS92]. For detailed accounts see surveys by Babai [B94] and Goldreich [G96].

Hastad's recent work [H96] contains a combinatorial consistency lemma which is related to our Lemma 1.1 (i.e., the "simple case" lemma). However, Hastad's lemma refers to the case where the test accepts with very low probability and so both its hypothesis and conclusion are weaker (and apparently harder to establish). Raz and Safra [RaSa] claim to have been inspired by our Lemma 1.2 (i.e., the "sparse case" lemma).

1.4 Organization

The (basic) "sparse case" consistency lemma is presented in Section 2. The application to the PCP Characterization Theorem is presented in Section 3. Section 4 contains a proof of Lemma 1.1 (which refers to sequences of totally independent random points).

Remark: This write-up reports work completed in the Spring of 1994, and announced at the Weizmann Workshop on Randomness and Computation (January 1995).

---

[4] Actually, [ALMSS] only prove agreement on an (arbitrary large) constant fraction of the domain.

2 The Consistency Lemma (for the sparse case)

In this section we present our main result; that is, a combinatorial consistency lemma which refers to sequences of bounded independence. Specifically, we considered k^2-long sequences viewed as k-by-k matrices. To emphasize the combinatorial nature of our lemma and its proof, we adopt an abstract presentation in which the properties required from the set of matrices are explicitly stated (as axioms). We comment that the set of all k-by-k matrices over S satisfies these axioms. A more important case is given in Construction 2.3: It is based on a standard construction of pairwise-independent sequences (i.e., the matrix is a pairwise-independent sequence of rows, where each row is a pairwise-independent sequence of elements).

General Notation. For a positive integer k, let $[k] \overset{\text{def}}{=} \{1, ..., k\}$. For a finite set A, the notation $a \in_R A$ means that a is uniformly selected in A. In case A is a multi-set, each element is selected with probability proportional to its multiplicity.

2.1 The Setting

Let S be some finite set, and let k be an integer. Both S and k are parameters, yet they will be implicit in all subsequent notations.

Rows and Columns. Let $\mathbf{R}$ be a multi-set of sequences of length k over S so that every $e \in S$ appears in some sequence of $\mathbf{R}$. For sake of simplicity, think of $\mathbf{R}$ as being a set (i.e., each sequence appears with multiplicity 1). Similarly, let $\mathbf{C}$ be another set of sequences (of length k over S). We neither assume $\mathbf{R} = \mathbf{C}$ nor $\mathbf{R} \neq \mathbf{C}$. We consider matrices having rows in $\mathbf{R}$ and columns in $\mathbf{C}$ (thus, we call the members of $\mathbf{R}$ row-sequences, and those in $\mathbf{C}$ column-sequences). We denote by $\mathbf{M}$ a multi-set of k-by-k matrices with rows in $\mathbf{R}$ and columns in $\mathbf{C}$. Namely,

Axiom 1 *For every $m \in \mathbf{M}$ and $i \in [k]$, the ith row of m is an element of $\mathbf{R}$ and the ith column of m is an element of $\mathbf{C}$.*

For every $i \in [k]$ and $\bar{r} \in \mathbf{R}$, we denote by $\mathbf{M}_i(\bar{r})$ the set of matrices (in $\mathbf{M}$) having $\bar{r}$ as the ith row. Similarly, for $j \in [k]$ and $\bar{c} \in \mathbf{C}$, we denote by $\mathbf{M}^j(\bar{c})$ the set of matrices (in $\mathbf{M}$) having $\bar{c}$ as the jth column. For every $\bar{r} = (r_1, ..., r_k) \in \mathbf{R}$ and every $\bar{c} = (c_1, ..., c_k) \in \mathbf{C}$, so that $r_j = c_i$, we denote by $\mathbf{M}_i^j(\bar{r}, \bar{c})$ the set of matrices having $\bar{r}$ as the ith row and $\bar{c}$ as the jth column (i.e., $\mathbf{M}_i^j(\bar{r}, \bar{c}) = \mathbf{M}_i(\bar{r}) \cap \mathbf{M}^j(\bar{c})$).

Shifts. We assume that $\mathbf{R}$ is "closed" under the shift operator. Namely,

Axiom 2 *For every $\bar{r} = (r_1, ..., r_k) \in \mathbf{R}$ there exists a unique $\bar{s} = (s_1, ..., s_k) \in \mathbf{R}$ satisfying $s_i = r_{i-1}$, for every $2 \leq i \leq k$. We denote this right-shifted sequence by $\sigma(\bar{r})$. Similarly, we assume that there exists a unique $\bar{s} = (s_1, ..., s_k) \in \mathbf{R}$*

satisfying $s_i = r_{i+1}$, for every $1 \leq i \leq k - 1$. We denote this left-shifted sequence by $\sigma^{-1}(\bar{r})$. Furthermore[5], we assume that shifting each of the rows of a matrix $m \in \mathbf{M}$, to the same direction, yields a matrix m' that is also in $\mathbf{M}$.

We stress that we do not assume that $\mathbf{C}$ is "closed" under shifts (in an analogous manner). For every (positive) integer i, the notations $\sigma^i(\bar{r})$ and $\sigma^{-i}(\bar{r})$ are defined in the natural way.

Distribution. We now turn to axioms concerning the distribution of rows and columns in a uniformly chosen matrix. We assume that the rows (and columns) of a uniformly chosen matrix are uniformly distributed in $\mathbf{R}$ (and $\mathbf{C}$, respectively).[6] In addition, we assume that the rows (but not necessarily the columns) are also pairwise independent. Specifically,

Axiom 3 *Let m be uniformly selected in $\mathbf{M}$. Then,*

1. *For every $i \in [k]$, the ith column of m is uniformly distributed in $\mathbf{C}$.*
2. *For every $i \in [k]$, the ith row of m is uniformly distributed in $\mathbf{R}$.*
3. *Furthermore, for every $j \neq i$ and $\bar{r} \in \mathbf{R}$, conditioned that the ith row of m equals $\bar{r}$, the jth row of m is uniformly distributed over $\mathbf{R}$.*

Finally, we assume that the columns in a uniformly chosen matrix containing a specific row-sequence are distributed identically to uniformly selected columns with the corresponding entry. A formal statement is indeed in place.

Axiom 4 *For every $i, j \in [k]$ and $\bar{r} = (r_1, ..., r_k) \in \mathbf{R}$, the jth column in a matrix that is uniformly selected among those having $\bar{r}$ as its ith row (i.e., $m \in_{\mathbf{R}} \mathbf{M}_i(\bar{r})$), is uniformly distributed among the column-sequences that have r_j as their i^{th} element.*

Clearly, if the jth element of $\bar{r} = (r_1, ..., r_k)$ differs from the ith element of $\bar{c} = (c_1, ..., c_k)$ then $\mathbf{M}_i^j(\bar{r}, \bar{c})$ is empty. Otherwise (i.e., $r_j = c_i$), by the above axiom, $\mathbf{M}_i^j(\bar{r}, \bar{c})$ is not empty.

2.2 The Test

Let Γ be a function assigning matrices in $\mathbf{M}$ (which may be a proper subset of all possible k-by-k matrices over S) values which are k-by-k matrices over some set of values V (i.e., $\Gamma : \mathbf{M} \mapsto V^{k \times k}$). The function Γ is *supposed* to be "consistent" (i.e., assign each element, e, of S the same value, independently of the matrix in which e appears). The purpose of the following test is to check that this property holds in some approximate sense.

Construction 2.1 (Consistency Test):

---

[5] The extra axiom is not really necessary; see remark following the definition of the consistency test.

[6] This, in fact, implies Axiom 1.

1. column test: *Select a column-sequence $\bar{c}$ uniformly in $\mathbf{C}$, and $i,j \in_R [k]$. Select two random extensions of this column, namely $m_1 \in_R \mathbf{M}^i(\bar{c})$ and $m_2 \in_R \mathbf{M}^j(\bar{c})$, and test if the ith column of $\Gamma(m_1)$ equals the jth column of $\Gamma(m_2)$.*

2. row test (analogous to the column test): *Select a row-sequence $\bar{r}$ uniformly in $\mathbf{R}$, and $i,j \in_R [k]$. Select two random extensions of this row, namely $m_1 \in_R \mathbf{M}_i(\bar{r})$ and $m_2 \in_R \mathbf{M}_j(\bar{r})$, and test if the ith row of $\Gamma(m_1)$ equals the jth row of $\Gamma(m_2)$.*

3. shift test: *Select a matrix m uniformly in $\mathbf{M}$ and an integer $t \in [k-1]$. Let m' be the matrix obtained from m by shifting each row by t; namely, the ith row of m' is $\sigma^t(\bar{r})$, where $\bar{r}$ denotes the ith row of m. We test if the $k-t$ first columns of $\Gamma(m)$ match the $k-t$ last columns of $\Gamma(m')$.*

The test accepts if all three (sub-)tests succeed.

Remark: Actually, it suffices to use a seemingly weaker test in which the row-test and shift-test are combined into the following **generalized row-test**: Select a row-sequence $\bar{r}$ uniformly in $\mathbf{R}$, integers $i,j \in_R [k]$ and $t \in_R \{0,1,...,k-1\}$. Select a random extension of this row and its shift, namely $m_1 \in_R \mathbf{M}_i(\bar{r})$ and $m_2 \in_R \mathbf{M}_j(\sigma^t(\bar{r}))$, and test if the $(k-t)$-long suffix of the ith row of $\Gamma(m_1)$ equals the $(k-t)$-long prefix of the jth row of $\Gamma(m_2)$.

Our main result asserts that Construction 2.1 is a "good consistency test": Not only that ALMOST ALL ENTRIES *in almost all matrices* are assigned in a consistent manner (which would have been obvious), but ALL ENTRIES IN ALMOST ALL ROWS *of almost all matrices* are assigned in a consistent manner.

Lemma 2.2 *Suppose $\mathbf{M}$ satisfies Axioms 1–4. Then, for every constant $\delta > 0$, there exist a constant $\epsilon > 0$ so that if a function $\Gamma : \mathbf{M} \mapsto V^{k \times k}$ passes the consistency test with probability at least $1 - \epsilon$ then there exists a function $\tau : S \mapsto V$ so that, with probability at least $1 - \delta$, the value assigned by Γ to a uniformly chosen matrix matches the values assigned by τ to the elements of a uniformly chosen row in this matrix. Namely,*

$$\text{Prob}_{i,m}(\forall j \; : \; \Gamma(m)_{i,j} = \tau(m_{i,j})) \; \geq \; 1 - \delta$$

where $m \in_R \mathbf{M}$ and $i \in_R [k]$. The constant ϵ does not depend on k and S. Furthermore, it is polynomially related to δ.

As a corollary, we get Part (1) of Lemma 1.2. Part (2) follows from Proposition 2.4 (below).

2.3 Proof of Lemma 2.2

As a motivation towards the proof of Lemma 2.2, consider the following mental experiment. Let $m \in \mathbf{M}$ be an arbitrary matrix and e be its $(i,j)^{\text{th}}$ entry. First, uniformly select a random matrix, denoted m_1, containing the ith row of m. Next, uniformly select a random matrix, denoted m_2, containing the jth column of m_1. The claim is that m_2 is uniformly distributed among the matrices containing the element e. Thus, if Γ passes items (1) and (2) in the consistency test then it must assign consistent values to almost all elements in almost all matrices. Yet, this falls short of even proving that there exists an assignment which matches all values assigned to the elements of some row in some matrix. Indeed, consider a function Γ which assigns 0 to all elements in the first ϵk columns of each matrix and 1's to all other elements. Clearly, Γ passes the row-test with probability 1 and the column-test with probability greater than $1 - \epsilon$; yet, there is no $\tau : S \mapsto V$ so that for a random matrix the values assigned by Γ to some row match τ. It is easy to see that the shift-test takes care of this special counter-example. Furthermore, it may be telling to see what is wrong with some naive arguments. A main issue these arguments tend to ignore is that for an "adversarial" choice of Γ and a candidate choice of $\tau : S \mapsto V$, we have no handle on the (column) *location* of the elements in a random matrix on which τ disagrees with Γ. The shift-test plays a central role in getting around this problem; see subsection 2.3 and Claim 2.2.14 (below).

Comment: The proofs of all claims are given in our technical report [GS96]. We believe that all the claims are quite believable, and that their proofs (though slightly tedious in some cases) are quite straightforward. In contrast, we believe that the ideas underlying the proof of the lemma are to be found in its high level structure; namely, the definitions and the claims made.

Notation: The following notation will be used extensively throughout the proof. For a k-by-k matrix, m, we denote by $\text{row}_i(m)$ the ith row of m and by $\text{col}^j(m)$ the jth column of m. Finally, denoting by $\text{entry}_{i,j}(m)$ the $(i, j)^{\text{th}}$ entry in the matrix m, we restate the conclusion of the lemma as follows

$$\text{Prob}_{i,m}(\exists j \text{ so that } \text{entry}_{i,j}(\Gamma(m)) \neq \tau(\text{entry}_{i,j}(m))) \leq \delta \tag{1}$$

where $m \in_{\text{R}} \mathbf{M}$ and $i \in_{\text{R}} [k]$.

Stable Rows and Columns – Part 1 For each $\bar{r} \in \mathbf{R}$ and $\bar{\alpha} \in V^k$, we denote by $p_{\bar{r}}(\bar{\alpha})$ the probability that Γ assigns to the row-sequence $\bar{r}$ the value-sequence $\bar{\alpha}$; namely,

$$p_{\bar{r}}(\bar{\alpha}) \overset{\text{def}}{=} \text{Prob}_{i,m}(\text{row}_i(\Gamma(m)) = \bar{\alpha})$$

where $i \in_{\text{R}} [k]$ and $m \in_{\text{R}} \mathbf{M}_i(\bar{r})$. The Row-Test implies that for almost all row-sequences there is a "typical" sequence of values; see Claim 2.2.3 (below).

Definition 2.2.1 (consensus): *The consensus of a row-sequence $\bar{r} \in \mathbf{R}$, denoted $\text{con}(\bar{r})$, is defined as the value $\bar{\alpha}$ for which $p_{\bar{r}}(\bar{\alpha})$ is maximum. Namely, $\text{con}(\bar{r}) = \bar{\alpha}$ if $\bar{\alpha}$ is the (lexicographically first) value-sequence for which $p_{\bar{r}}(\bar{\alpha}) = \max_{\bar{\beta}}\{p_{\bar{r}}(\bar{\beta})\}$.*

Definition 2.2.2 (stable sequences): *Let $\epsilon_2 \overset{\text{def}}{=} \sqrt{\epsilon}$. We say that the row-sequence $\bar{r}$ is stable if $p_{\bar{r}}(\text{con}(\bar{r})) \geq 1 - \epsilon_2$. Otherwise, we say that $\bar{r}$ is unstable.*

Clearly, almost all row-sequences are stable. That is,

Claim 2.2.3 *All but at most an ϵ_2 fraction of the row-sequence are stable.*

By definition, almost all matrices containing a particular *stable* row-sequence assign this row-sequence the same sequence of values (i.e., its consensus value). We say that such matrices are conforming for this row-sequence.

Definition 2.2.4 (conforming matrix): *Let $i \in [k]$. A matrix $m \in M$ is called i-conforming (or conforming for row-position i) if Γ assigns the ith row of m its consensus value; namely, if $\mathrm{row}_i(\Gamma(m)) = \mathrm{con}(\mathrm{row}_i(m))$. Otherwise, the matrix is called i-non-conforming (or non-conforming for row-position i).*

Claim 2.2.5 *The probability that for a uniformly chosen $i \in [k]$ and $m \in M$, the matrix m is i-non-conforming is at most $\epsilon_3 \stackrel{\text{def}}{=} 2\epsilon_2$. Furthermore, the bound holds also if we require that the ith row of m is stable.*

Remark: Clearly, an analogous treatment can be applied to column-sequences. In the sequel, we freely refer to the above notions and to the above claims also when discussing column-sequences.

Stable Rows – Part 2 (Shifts) Now we consider the relation between the consensus of row-sequences and the consensus of their (short) shifts. By a short shift of the row-sequence $\bar{r}$, we mean any row-sequence $\bar{s} = \sigma^d(\bar{r})$ obtained with $d \in \{-(k-1), ..., +(k-1)\}$. Our aim is to show that the consensus (as well as stability) is usually preserved under short shifts.

Definition 2.2.6 (very-stable row): *Let $\epsilon_4 = \sqrt{\epsilon_2}$. We say that a row-sequence $\bar{r}$ is very stable if it is stable, and for all but an ϵ_4 fraction of $d \in \{-(k-1), ..., +(k-1)\}$, the row-sequence $\bar{s} \stackrel{\text{def}}{=} \sigma^d(\bar{r})$ is also stable.*

Clearly,

Claim 2.2.7 *All but at most an ϵ_4 fraction of the row-sequence are very-stable.*

Definition 2.2.8 (super-stable row): *Let $\epsilon_5 = \sqrt[3]{\epsilon}$ and $\epsilon_6 = 2(\epsilon_4 + \epsilon_5)$. We say that a row-sequence $\bar{r}$ is super-stable if it is very-stable, and, for every $j \in [k]$, the following holds*

> *for all but an ϵ_6 fraction of the $t \in [k]$, the row-sequence $\bar{s} \stackrel{\text{def}}{=} \sigma^{t-j}(\bar{r})$ is stable and $\mathrm{con}_j(\bar{r}) = \mathrm{con}_t(\bar{s})$, where $\mathrm{con}_j(\bar{r})$ is the jth element of $\mathrm{con}(\bar{r})$.*

Note that the tth element of $\sigma^{t-j}(\bar{r})$ is $r_{t-(t-j)} = r_j$. Thus, a row-sequence is super-stable if the consensus value of each of its elements is preserved under almost all (short) shifts.

Claim 2.2.9 *All but at most an ϵ_6 fraction of the row-sequence are super-stable.*

Deriving the Conclusion of the Lemma We are now ready to derive the conclusion of the Lemma. Loosely speaking, we claim that the function τ, defined so that $\tau(e)$ is the value most frequently assigned (by Γ) to e, satisfies Eq. (1). Actually, we use a slightly different definition for the function τ.

Definition 2.2.10 (the function τ): *For a column-sequence $\bar{c}$, we denote by $con_i(\bar{c})$ the values that $con(\bar{c})$ assigns to the ith element in $\bar{c}$. We denote by $\mathbf{C}_i(e)$ the set of column-sequences having e as the ith component. Let $q_e(v)$ denote the probability that the consensus of a uniformly chosen column-sequence, containing e, assigns to e the value v. Namely,*

$$q_e(v) \overset{\text{def}}{=} \text{Prob}_{i,\bar{c}}(con_i(\bar{c}) = v)$$

where $i \in_R [k]$ and $\bar{c} \in_R \mathbf{C}_i(e)$. We consider $\tau : S \mapsto V$ so that $\tau(e) \overset{\text{def}}{=} v$ if $q_e(v) = \max_u\{q_e(u)\}$, with ties broken arbitrarily.

Assume, on the contrary to our claim, that Eq. (1) does not hold (for this τ). Namely, for a uniformly chosen $m \in \mathbf{M}$ and $i \in [k]$, the following holds with probability greater that δ

$$\exists j \text{ so that } \text{entry}_{i,j}(\Gamma(m)) \neq \tau(\text{entry}_{i,j}(m)) \tag{2}$$

The notion of a annoying row-sequence, defined below, plays a central role in our argument. Using the above (contradiction) hypothesis, we first show that many row-sequences are annoying. Next, we show that lower bounds on the number of annoying row-sequences translate to lower bounds on the probability that a uniformly chosen matrix is non-conforming for a uniformly chosen column position. This yields a contradiction to Claim 2.2.5.

Definition 2.2.11 (row-annoying elements): *An element r_j in $\bar{r} = (r_1, ..., r_k) \in \mathbf{R}$, is said to be **annoying for the row-sequence** $\bar{r}$ if the jth element in $con(\bar{r})$ differs from $\tau(r_j)$. A row-sequence $\bar{r}$ is said to be **annoying** if $\bar{r}$ contains an element that is annoying for it.*

Using Claim 2.2.9, we get

Claim 2.2.12 *Suppose that Eq. (1) does not hold (for τ). Then, at least a $\delta_1 \overset{\text{def}}{=} \delta - \epsilon_6 - \epsilon_2$ fraction of the row-sequences are both super-stable and annoying.*

A key observation is that each stable row-sequence which is annoying yields many matrices which are non-conforming for the "annoying column position" (i.e., for the column position containing the element which annoys this row-sequence). Namely,

Claim 2.2.13 *Suppose that a row-sequence $\bar{r} = (r_1, ..., r_k)$ is stable and that r_j is annoying for $\bar{r}$. Then, at least a $\frac{1}{2} - \epsilon_2$ fraction of the matrices, containing the row-sequence $\bar{r}$, are non-conforming for their j^{th} column.*

We stress that the row-sequence $\bar{r}$ in the above claim is *not* necessarily very-stable (let alone super-stable). Another key observation is that super-stable row-sequences which are annoying have the property of "infecting" almost all their shifts with their annoying positions, and thus spreading the "annoyance" over all column positions. Namely,

Claim 2.2.14 *Suppose that a row-sequence $\bar{r}$ is both super-stable and annoying. In particular, suppose that the jth element of $\bar{r} = (r_1, ..., r_k)$ is annoying for $\bar{r}$. Then, for all but at most an ϵ_6 fraction of the $t \in [k]$, the the row-sequence $\bar{s} = \sigma^{t-j}(\bar{r})$ is stable and its tth element (which is indeed r_j) is annoying for $\bar{s}$.*

Combining Claims 2.2.12 and 2.2.14, we derive, for almost all positions $t \in [k]$, a lower bound for the number of stable row-sequences that are annoyed by their t^{th} element.

Claim 2.2.15 *Suppose that Eq. (1) does not hold (for τ). Then, there exists a set $T \subseteq [k]$ so that $|T| \geq (1 - 2\epsilon_6) \cdot k$ and for every $t \in T$ there is a set of at least $\frac{\delta_1}{2k} \cdot |\mathbf{R}|$ stable row-sequences so that the tth position is annoying for each of these sequences.*

Combining Claims 2.2.15 and 2.2.13, we get a lower bound on the number of matrices which are non-conforming for the jth column, $\forall j \in T$ (where T is as in Claim 2.2.15). Namely,

Claim 2.2.16 *Let T be as guaranteed by Claim 2.2.15 and suppose that $j \in T$. Then, at least a $\frac{\delta_1}{6}$ fraction of the matrices are non-conforming for column-position j.*

The combination of Claims 2.2.15 and 2.2.16, yields that a uniformly chosen matrix is non-conforming for a uniformly chosen column position with probability at least $(1 - 2\epsilon_6) \cdot \frac{\delta_1}{6}$. For a suitable choice of constants (e.g., $\epsilon = (\delta/30)^4$), this yields contradiction to Claim 2.2.5. Thus, Eq. (1) must hold for τ as defined in Def. 2.2.10, and the lemma follows. ■

2.4 A Construction that Satisfies the Axioms

Clearly, the set of all k-by-k matrices over S satisfies Axioms 1–4. A more interesting and useful set of matrices is defined as follows.

Construction 2.3 (basic construction): *We associate the set S with a finite field and suppose $k \leq |S|$. Furthermore, $[k]$ is associated with k elements of the field so that 1 is the multiplicative unit and $i \in [k]$ is the sum of i such units. Let $\mathbf{M}$ be the set of matrices defined by four field elements as follows. The matrix associated with the quadruple (x, y, x', y') has the $(i, j)^{\text{th}}$ entry equal $(x + jy) + i(x' + jy')$.*

Remark: The column-sequences correspond to the standard pairwise-independent sequences $\{r + is : i \in [k]\}$, where $r, s \in S$. Similarly, the row-sequences are expressed as $\{r + js : j \in [k]\}$, where $r, s \in S$.

Proposition 2.4 *The Basic Construction satisfies Axioms 1-4.*

3 A Stronger Consistency Test and the PCP Application

To prove Lemma 1.3, we need a slightly stronger consistency test than the one analyzed in Lemma 2.2. This new test is given access to three related oracles, each supplying assignments to certain classes of sequences over S, and is supposed to establish the consistency of these oracles with one function $\tau : S \mapsto V$. Specifically, one oracle assigns values to k^2-long sequences viewed as two-dimensional arrays (as before). The other two oracles assign values to k^3-long sequences viewed as 3-dimensional arrays, whose slices (along a specific coordinate) correspond to the 2-dimensional arrays of the first oracle. Using Lemma 2.2 (and the auxiliary oracles) we will present a test which verifies that the first oracle is consistent in an even stronger sense than established in Lemma 2.2. Namely, not only that *all entries* IN ALMOST ALL ROWS *of almost all 2-dimensional arrays* are assigned in a consistent manner, but ALL ENTRIES *in almost all 2-dimensional arrays are assigned in a consistent manner.*

3.1 The Setting

Let S, k, $\mathbf{R}$, $\mathbf{C}$ and $\mathbf{M}$ be as in the previous section. We now consider a family, $\mathcal{M}_c$, of k-by-k matrices with entries is $\mathbf{C}$. The family $\mathcal{M}_c$ will satisfy Axioms 1–4 of the previous section. In addition, its induced multi-set of row-sequences, denoted $\mathcal{R}$, will correspond to the multi-set $\mathbf{M}$; namely, each row of a matrix in $\mathcal{M}_c$ will form a matrix in $\mathbf{M}$ (i.e., the sequence of elements of $\mathbf{C}$ corresponding to a row in a $\mathcal{M}_c$-matrix will correspond to a $\mathbf{M}$-matrix). Put formally,

Axiom 5 *For every* $\mathbf{m} \in \mathcal{M}_c$ *and every* $i \in [k]$, *there exists* $m \in \mathbf{M}$ *so that for every* $j \in [k]$, *the* $(i,j)^{\text{th}}$ *entry of* $\mathbf{m}$ *equals the* j^{th} *column of* m *(i.e.,* $\text{entry}_{i,j}(\mathbf{m}) = \text{col}^j(m)$, *or, equivalently,* $\text{row}_i(\mathbf{m}) \cong m$). *Furthermore, this matrix* m *is unique.*

Analogously, we consider also a family, $\mathcal{M}_r$, of k-by-k matrices the entries of which are elements in $\mathbf{R}$ so that the rows[7] of each $\mathbf{m} \in \mathcal{M}_r$ correspond to matrices in $\mathbf{M}$.

3.2 The Test

As before, Γ is a function assigning (k-by-k) matrices in $\mathbf{M}$ values which are k-by-k matrices over some set of values V (i.e., $\Gamma : \mathbf{M} \mapsto V^{k \times k}$). Let Γ_c (resp., Γ_r) be (the supposedly corresponding) function assigning k-by-k matrices over $\mathbf{C}$ (resp., $\mathbf{R}$) values which are k-by-k matrices over $\overline{V} \stackrel{\text{def}}{=} V^k$ (i.e., $\Gamma_c : \mathcal{M}_c \mapsto \overline{V}^{k \times k}$).

Construction 3.1 (Extended Consistency Test):

---

[7] Alternatively, one can consider a family, $\mathcal{M}_r$, of k-by-k matrices the entries of which are elements in $\mathbf{R}$ so that the columns of each $\mathbf{m} \in \mathcal{M}_r$ correspond to matrices in $\mathbf{M}$. However, this would require to modify the basic consistency test (of Construction 2.1), for these matrices, so that it shifts columns instead of rows.

1. consistency for sequences: *Apply the consistency test of Construction 2.1 to Γ_c. Same for Γ_r.*

2. correspondence test: *Uniformly select a matrix $\mathbf{m} \in \mathcal{M}_c$ and a row $i \in [k]$, and compare the ith row in $\Gamma_c(\mathbf{m})$ to $\Gamma(m)$, where $m \in \mathbf{M}$ is the matrix formed by the $\mathbf{C}$-elements in the ith row of $\mathbf{m}$. Same for Γ_r.*

The test accepts if both (sub-)tests succeed.

Lemma 3.2 *Suppose $\mathbf{M}, \mathcal{M}_c, \mathcal{M}_r$ satisfy Axioms 1–5. Then, for every constant $\gamma > 0$, there exist a constant ϵ so that if a function $\Gamma : \mathbf{M} \mapsto V^{k \times k}$ (together with some functions $\Gamma_c : \mathcal{M}_c \mapsto \overline{V}^{k \times k}$ and $\Gamma_r : \mathcal{M}_r \mapsto \overline{V}^{k \times k}$) passes the extended consistency test with probability at least $1 - \epsilon$ then there exists a function $\tau : S \mapsto V$ so that, with probability at least $1 - \gamma$, the value assigned by Γ to a uniformly chosen matrix $m \in \mathbf{M}$ matches the values assigned by τ to each of the elements of m. Namely,*

$$\mathrm{Prob}_m \left(\forall i, j \ \mathrm{entry}_{i,j}(\Gamma(m)) = \tau(\mathrm{entry}_{i,j}(m)) \right) \geq 1 - \gamma$$

where $m \in_R \mathbf{M}$. The constant ϵ does not depend on k and S. Furthermore, it is polynomially related to γ.

The proof of the lemma starts by applying Lemma 2.2 to derive assignments to $\mathbf{C}$ (resp., $\mathbf{R}$) which are consistent with Γ_c (resp., Γ_r) on almost all rows of almost all k^3-dimensional arrays (i.e., $\mathcal{M}_c$ and $\mathcal{M}_r$, respectively). It proceeds by applying a degenerate argument of the kind applied in the proof of Lemma 2.2. For details see our technical report [GS96].

3.3 Application to Low-Degree Testing

Again, the set of all k-by-k-by-k arrays over S satisfies Axioms 1–5. A more useful set of 3-dimensional arrays is defined as follows.

Construction 3.3 (main construction): *Let $\mathbf{M}$ be as in the Basic Construction (i.e., Construction 2.3). We let $\mathcal{M}_c = \mathcal{M}_r$ be the set of matrices defined by applying the Basic Construction to the element-set $\mathbf{C} = \mathbf{R}$. Specifically, a matrix in $\mathcal{M}_c$ is defined by the quadruple (x, y, x', y'), where each of the four elements is a pair over S, so that the $(i, j)^{th}$ entry in the matrix equals $(x + jy) + i(x' + jy')$. Here x, y, x', y' are viewed as two-dimensional vectors over the finite field S and i, j are scalars in S. The $(i, j)^{th}$ entry is a pair over S which represents a pairwise independent sequence (which equals an element in $\mathbf{C} = \mathbf{R}$).*

Clearly,

Claim 3.4 *Construction 3.3 satisfies Axioms 1–5.*

Combining all the above with the low-degree test of [GLRSW, RS96] using the results claimed there[8], we get a low-degree test which is sufficiently efficient to be used in the proof of the PCP-Characterization of NP.

---

[8] Rather than using much stronger results obtained via a more complicated analysis, as in [ALMSS], which rely on the Lemma of [AS].

Construction 3.5 (Low Degree Test): *Let $f : F^n \mapsto F$, where F is a field of prime cardinality, and d be an integer so that $|F| > 4(d+2)^2$. Let $\mathbf{M}$, $\mathcal{M}_c$ and $\mathcal{M}_r$ be as in Construction 3.3, with $S = F^n$, $V = F$ and $k \stackrel{\text{def}}{=} 4(d+2)^2$. Let $\Gamma : \mathbf{M} \mapsto F^{k \times k}$, $\Gamma_r : \mathcal{M}_r \mapsto F^{k^3}$ and $\Gamma_c : \mathcal{M}_c \mapsto F^{k^3}$ be auxiliary tables (which should contain the corresponding f-values). The low degree test consists of*

1. *Applying the Extended Consistency Test to $\Gamma : \mathbf{M} \mapsto F^{k \times k}$, $\Gamma_r : \mathcal{M}_r \mapsto F^{k^3}$ and $\Gamma_c : \mathcal{M}_c \mapsto F^{k^3}$.*
2. *Selecting uniformly a matrix $m \in \mathbf{M}$ and testing that the Polynomial Interpolation Condition (cf., [GLRSW]) holds for each row; namely, we test that*

$$\sum_{i=1}^{d+2} \alpha_i \cdot \text{entry}_{i,j}(\Gamma(m)) = 0$$

   for all $j \in [k]$, where $\alpha_i = (-1)^i \cdot \binom{d+1}{i-1}$.
3. *Select uniformly a matrix in $\mathbf{M}$ and test matching of random entry to f. Namely, select uniformly $m \in \mathbf{M}$, and $i, j \in [k]$, and check if $\text{entry}_{i,j}(\Gamma(m)) = f(\text{entry}_{i,j}(m))$.*

The test accepts if and only if all the above three sub-tests accept.

Proposition 3.6 *Let $f : F^n \mapsto F$, where F is a prime field, and let $\ell \stackrel{\text{def}}{=} n \cdot \log_2 |F|$. Then, the Low Degree Test of Construction 3.5 requires $O(\ell)$ randomness and query length, $\text{poly}(\ell)$ answer length and satisfies:*

completeness: *If f is a degree-d polynomial, then there exist $\Gamma : \mathbf{M} \mapsto F^{k \times k}$, $\Gamma_r : \mathcal{M}_r \mapsto F^{k^3}$ and $\Gamma_c : \mathcal{M}_c \mapsto F^{k^3}$ so that the test always accepts.*

soundness: *For every $\delta > 3/(d+2)^2$ there exists an $\epsilon > 0$ so that for every f which is at distance at least δ from any degree-d polynomial and for every $\Gamma : \mathbf{M} \mapsto F^{k \times k}$, $\Gamma_r : \mathcal{M}_r \mapsto F^{k^3}$ and $\Gamma_c : \mathcal{M}_c \mapsto F^{k^3}$, the test rejects with probability at least ϵ. Furthermore, the constant ϵ is a polynomial in δ which does not depend on n, d and F.*

As a corollary, we get Lemma 1.3.

Proof: As usual, the completeness clause is easy to establish. We thus turn to the soundness requirement. By Claim 3.4, we may apply Lemma 3.2 to the first sub-test and infer that either the first sub-test fails with some constant probability (say ϵ_1) or there exists a function $\tau : F^n \mapsto F$ so that with very high constant probability (say $1 - \delta_1$)

$$\text{entry}_{i,j}(\Gamma(m)) = \tau(\text{entry}_{i,j}(m)) \tag{3}$$

holds for all $i \in [d+2]$ and $j \in [k]$. On the other hand, by [GLRSW] (see also [S95, Thm 3.3] and [RS96, Thm 5]), either

$$\text{Prob}_{x,y \in F^n} \left(\sum_{i=1}^{d+2} \alpha_i \cdot \tau(x + iy) \neq 0 \right) > \frac{1}{2(d+2)^2} \tag{4}$$

or τ is very close (specifically at distance at most $1/(d+2)^2$) to some degree-d polynomial. A key observation is that the Main Construction (i.e., Construction 3.3) has the property that rows in $m \in_R \mathbf{M}$ are distributed identically to the distribution in Eq. (4). Thus, for every $j \in [k]$ either

$$\text{Prob}_{m \in \mathbf{M}} \left(\sum_{i=1}^{d+2} \alpha_i \cdot \tau(\text{entry}_{i,j}(m)) \neq 0 \right) > \frac{1}{2(d+2)^2} \tag{5}$$

or τ is at distance at most $\delta_2 \stackrel{\text{def}}{=} 1/(d+2)^2$ from some degree-d polynomial. However, we claim that in case Eq. (5) holds, the second sub-test will reject with constant probability. The claim is proven by first considering $k = 4(d+2)^2$ copies of the GLRSW Test (i.e., the test in Eq. (5)). Using Chebishev's Inequality and the hypothesis by which each copy rejects with probability at least $1/2(d+2)^2$, we conclude that the probability that none of these copies rejects is bounded above by $\frac{2(d+2)^2}{4(d+2)^2} = \frac{1}{2}$. Thus, the second sub-test must reject with probability at least $\epsilon_2 \stackrel{\text{def}}{=} \frac{1}{2} - \delta_1$, where δ_1 accounts for the substitution of the τ values by the entries in $\Gamma(\cdot)$. We conclude that τ must be δ_2-close to a degree-d polynomial or else the test rejects with too high probability (i.e., ϵ_2). Finally, we claim that if f disagrees with τ on $\delta_3 > \delta_1$ of the inputs then the third sub-test rejects with probability at least $\epsilon_3 \stackrel{\text{def}}{=} \delta_3 - \delta_1$ (since the distance from f to τ is bounded by the sum of the distances of f to the matrix and of τ to the matrix). The proposition follows using some arithmetics: Specifically, we set $\delta_1 = \delta/3$, $\delta_3 = 2\delta_1$, $\epsilon_1 = \text{poly}(\delta_1)$ (as in Lemma 3.2), and verify that $\delta_3 + \delta_2 \leq \delta$ (since $\delta/3 \geq (d+2)^{-2} = \delta_2$), and $\epsilon = \min\{\epsilon_1, \epsilon_2, \epsilon_3\} = \text{poly}(\delta)$ (since $\epsilon_2 \geq 1/6$ and $\epsilon_3 = \delta/3$). ∎

4 Proof of Lemma 1.1

Having proven Lemma 2.2, we can apply it[9] to derive a short proof of Lemma 1.1. To this end we view ℓ-multisets over S as k-by-k matrices, where $k = \sqrt{\ell}$. Recall that the resulting set of matrices satisfies Axioms 1–4. Thus, by Lemma 2.2, in case the test accepts with probability at least $1 - \epsilon$, there exists a function $\tau : S \mapsto V$ such that

$$\text{Prob}_{A \in_R S^k, B \in_R E_{k^2}(A)}(\forall e \in A, \ \Gamma(B)_e = \tau(e)) \geq 1 - \delta$$

where S^k is the set of all k-multisets over S and $E_l(A)$ is the set of all l-multisets extending A. We can think of this probability space as first selecting $B \in_R S^{k^2}$ and next selecting a k-subset A in B. Thus,

$$\text{Prob}_{B \in_R S^{k^2}, A \in_R C_k(B)}(\exists e \in A \text{ s.t. } \Gamma(B)_e \neq \tau(e)) \leq \delta \tag{6}$$

---

[9] This is indeed an over-kill. For example, we can avoid all complications regarding shifts (in the proof of Lemma 2.2).

where $C_k(B)$ denotes the set of all k-multisets contained in B. This implies

$$\text{Prob}_{B \in_R S^{k^2}} (|\{e \in B : \Gamma(B)_e \neq \tau(e)\}| > k) \leq 2\delta$$

as otherwise Eq. (6) is violated. (The probability that a random k-subset hits a subset of density $\frac{1}{k}$ is at least $\frac{1}{2}$.) The lemma follows.

Acknowledgment

Research was supported in part by grant No. 92-00226 from the United States – Israel Binational Science Foundation (BSF), Jerusalem, Israel. We are grateful to Madhu Sudan for pointing out an error in an earlier version, and for other helpful comments.

References

[ECCC] The *Electronic Colloquium on Computational Complexity*, ECCC, is a new forum for the rapid and widespread interchange of ideas in computational complexity. Online access to ECCC is possible using http://www.eccc.uni-trier.de/eccc/.

[ALMSS] S. Arora, C. Lund, R. Motwani, M. Sudan and M. Szegedy. Proof Verification and Intractability of Approximation Problems. In *33rd FOCS*, pages 14–23, 1992.

[AS] S. Arora and S. Safra. Probabilistic Checkable Proofs: A New Characterization of NP. In *33rd FOCS*, pages 1–13, 1992.

[B94] L. Babai. Transparent Proofs and Limits to Approximation. TR-94-07, Dept. of Computer Science, University of Chicago, 1994.

[BFL] L. Babai, L. Fortnow, and C. Lund. Non-Deterministic Exponential Time has Two-Prover Interactive Protocols. In *31st FOCS*, pages 16–25, 1990.

[BFLS] L. Babai, L. Fortnow, L. Levin, and M. Szegedy. Checking Computations in Polylogarithmic Time. In *23rd STOC*, pages 21–31, 1991.

[BF] D. Beaver and J. Feigenbaum. Hiding Instances in Multioracle Queries. In *7th STACS*, Springer Verlag, LNCS Vol. 415, pages 37–48, 1990.

[BGS] M. Bellare, O. Goldreich and M. Sudan. Free Bits, PCPs and Non-Approximability – Towards Tight Results. Extended abstract in *36th FOCS*, pages 422–431, 1995. Full version appears in ECCC TR95-024. 1995.

[BGLR] M. Bellare, S. Goldwasser, C. Lund and A. Russell. Efficient Probabilistically Checkable Proofs and Applications to Approximation. In *25th STOC*, pages 294–304, 1993.

[BS] M. Bellare and M. Sudan. Improved Non-Approximability Results. In *26th STOC*, pages 184–193, 1994.

[BGKW] M. Ben-Or, S. Goldwasser, J. Kilian and A. Wigderson. Multi-Prover Interactive Proofs: How to Remove Intractability. In *20th STOC*, pages 113–131, 1988.

[BLR] M. Blum, M. Luby and R. Rubinfeld. Self-Testing/Correcting with Applications to Numerical Problems. In *22nd STOC*, 1990.

[FGLSS] U. Feige, S. Goldwasser, L. Lovász, S. Safra, and M. Szegedy. Approximating Clique is almost NP-complete. In *32nd FOCS*, pages 2–12, 1991.

[FRS]   L. Fortnow, J. Rompel and M. Sipser. On the Power of Multi-Prover Interactive Protocols. In *Proc. 3rd IEEE Symp. on Structure in Complexity Theory*, pages 156–161, 1988.

[FS]   K. Friedl and M. Sudan. Some Improvement to Total Degree Tests. In *Proc. 3rd Israel Symp. on Theory of Computing and Systems*, pages 190–198, 1995.

[GLRSW] P. Gemmell, R. Lipton, R. Rubinfeld, M. Sudan, and A. Wigderson. Self-Testing/Correcting for Polynomials and for Approximate Functions. In *23th STOC*, pages 32–42, 1991.

[G96]   O. Goldreich. A Taxonomy of Proof Systems. To appear in *Complexity Theory   Retrospective   II*, L.A. Hemaspaandra and A. Selman (eds.), Springer-Verlag, 1996. A copy is available from URL http://theory.lcs.mit.edu/~oded/pps.html.

[GMW]   O. Goldreich, S. Micali and A. Wigderson. Proofs that Yield Nothing but their Validity or All Languages in NP Have Zero-Knowledge Proof Systems. Extended abstract in *27th FOCS*, 1986.

[GS96]   O. Goldreich and M. Safra. A Combinatorial Consistency Lemma with application to proving the PCP Theorem. In **ECCC** TR96-047. Revised October 1996.

[GMR]   S. Goldwasser, S. Micali and C. Rackoff. The Knowledge Complexity of Interactive Proof Systems. *SIAM Journal on Computing*, Vol. 18, pages 186–208, 1989.

[H96]   J. Hastad. Clique is Hard to Approximate within $n^{1-\epsilon}$. In *37th FOCS*, pages 627–636, 1996.

[LS]   D. Lapidot and A. Shamir. Fully Parallelized Multi Prover Protocols for NEXP-time. In *32nd FOCS*, pages 13–18, 1991.

[LFKN]   C. Lund, L. Fortnow, H. Karloff, and N. Nisan. Algebraic Methods for Interactive Proof Systems. In *31st FOCS*, pages 2–10, 1990.

[RaSa]   R. Raz and S. Safra. A Sub-Constant Error-Probability Low-Degree Test and a Sub-Constant Error-Probability PCP Characterization of NP. In *29th STOC*, pages 475–484, 1997.

[RS92]   R. Rubinfeld and M. Sudan. Testing Polynomial Functions Efficiently and over Rational Domains. In *3rd SODA*, pages 23–32, 1992.

[RS96]   R. Rubinfeld and M. Sudan. Robust Characterization of Polynomials with Application to Program Testing. *SIAM J. of Computing*, Vol. 25, No. 2, pages 252–271, 1996. This paper is considered the journal version of [GLRSW, RS92].

[S90]   A. Shamir. IP=PSPACE. In *31st FOCS*, pages 11–15, 1990.

[S95]   M. Sudan. *Efficient Checking of Polynomials and Proofs and the Hardness of Approximation Problems*. ACM Distinguished Theses, Springer-Verlag, LNCS Vol. 1001, 1995.

Super-bits, Demi-bits, and $N\tilde{P}/qpoly$-natural Proofs

Steven Rudich

Computer Science Department, Carnegie Mellon University, Pittsburgh, PA 15213
rudich@cs.cmu.edu

Abstract. We introduce the super-bit conjecture, which allows the development of a theory generalizing the notion of pseudorandomness so as to fool *non-deterministic* statistical tests. This new kind of pseudorandomness rules out the existence of $NP/poly$-natural properties that can work against $P/poly$. This is an important extension of the original theory of $P/poly$-natural proofs [10]. We also introduce the closely related demi-bit conjecture which is more intuitive and is the source of interesting open problems.

1 Introduction

By exploiting the theory of pseudorandom generators [1, 11, 3, 7], Razborov and Rudich [10] give evidence that the proof techniques used in the last sixteen years of non-uniform, non-monotone circuit lower bounds will be unable to solve the barrier problems of complexity theory. They argue that known lower bound arguments are all "natural" which means that they exploit some P-natural combinatorial property (see Section 3.1). Furthermore, they show that if strong enough pseudorandomness is possible, no such property can be used in proving that satisfiability requires super-polynomial size circuits. Subsequently, Razborov [8, 9] showed that proof systems with the constructive interpolation property (see also [6]) cannot prove such lower bounds without implying the existence of the arguments ruled out in [10] under a pseudorandomness assumption. Thus, the existence of pseudorandom generators implies the independence of $SAT \notin P/poly$ from a class of proof systems.

We extend the standard notion of pseudorandomness so as to be secure against *non-deterministic* adversaries. (This is fairly counter-intuitive. A non-deterministic adversary can simply guess the seed of the generator.) Our extension makes the generalization of previous work quite easy. We introduce the super-bit conjecture, which serves as the hardness assumption to found our theory. We show that the super-bit conjecture rules out $NP/poly$ (in fact, $NquasiP/qpoly$) natural proofs that satisfiability requires large circuits. This is a much stronger negative result than the previous [10]. Furthermore, the super-bit conjecture implies the independence of $SAT \notin P/poly$ from any proof system with the *existential* interpolation property.

For concreteness, we propose a super-bit generator based on the non-deterministic hardness of the subset sum problem. Thus all the conjectures in this paper are true if subset sum is suitably hard.

We introduce the demi-bit conjecture, which is more intuitive than the super-bit conjecture. We hope, but we cannot prove, that this will serve as the foundation for our theory. The open problems about demi-hardness are of independent interest.

2 Definitions and Notation

We denote by F_n the set of all Boolean functions in n variables. Most of the time, it will be convenient to think of $f_n \in F_n$ as a binary string of length 2^n, called the *truth-table* of f_n.

A non-deterministic circuit is one which includes gates outputting a single "non-deterministic" bit. The circuit is said to accept iff there is a choice of non-deterministic bits to be output that causes the circuit to accept its input. $N\tilde{P}/qpoly$ is the notation for non-uniform, non-deterministic circuits of quasi-polynomial size.

3 The $N\tilde{P}/qpoly$-natural properties conjecture

3.1 Natural properties in general.

Razborov and Rudich [10], formalize the class of combinatorial properties that arise in published lower bound arguments in boolean circuit complexity. We start by reviewing their definition.

Let Γ and Λ be complexity classes. Call a combinatorial property C_n Γ-*natural* with density δ_n if it contains $C_n^* \subseteq C_n$ with the following two conditions:

Constructivity: The predicate $f_n \overset{?}{\in} C_n^*$ is computable in Γ (recall, C_n^* is a set of truth-tables with 2^n bits);
Largeness: $|C_n^*| \geq \delta_n \cdot |F_n|$.

A combinatorial property C_n is *useful against* Λ if it satisfies:

Usefulness: For any sequence of functions f_n, where the event $f_n \in C_n$ happens infinitely often, $\{f_n\} \notin \Lambda$.

3.2 $NP/poly$-natural properties.

Of interest here will be a choice of specific parameters in the general definition. Let $\Gamma = NP/poly$ and δ_n be any function of growth rate $2^{-O(n)}$.

Call a combinatorial property C_n $NP/poly$-*natural* if it contains $C_n^* \subseteq C_n$ with the following two conditions:

Constructivity: $C_n^* \in NP/poly$.
Largeness: $|C_n^*| \geq 2^{-O(n)} \cdot |F_n|$.

3.3 $N\tilde{P}/qpoly$-natural properties.

Also of interest will be a choice of parameters in the general definition with even more liberal parameters. Let $N\tilde{P}/qpoly$ be the class of languages recognized by non-uniform, quasi-polynomial-size circuit families. (Quasi-polynomial means $n^{\log n^{O(1)}}$.)

Let $\Gamma = N\tilde{P}/qpoly$ and δ_n be any function of growth rate $2^{-n^{O(1)}}$. If C_n is natural with these parameters, we say that it is $N\tilde{P}/qpoly$-natural.

3.4 Main conjecture

Conjecture 1 *There are no $N\tilde{P}/qpoly$-natural properties that are useful against $P/poly$.*

A trivial corollary is the weaker statement: *there are no $NP/poly$-natural properties useful against $P/poly$.*

4 The significance of the conjecture.

4.1 A pleasing overall concept.

Conjecture 1 is a much stronger statement than the main negative statement of [10] that there are no $P/poly$-natural proofs. It rules out a wider category of lower bound proof than we have yet encountered.

The concept of an $N\tilde{P}/qpoly$-natural proof is actually simpler than $P/poly$-natural proofs. $N\tilde{P}/qpoly$-natural proofs are any argument that gives a non-negligible fraction of boolean functions a short certificate proving that they lie outside some complexity class.

4.2 Relation to independence results

We say that a propositional proof system (see [5]) T has the *Existential Interpolation Property (EIP)* if there is a polynomial $p(x)$ such that whenever T proves $A(a_1, \ldots, a_n) \vee B(b_1, \ldots b_m)$ with a proof of length l then T has a $p(l)$ length proof of either A or B. (Notice that the a_i's and b_j's are distinct variables.)

We say that a propositional proof system T has the *Constructive Interpolation Property (CIP)* if it has the EIP and there is a polynomial time procedure that, given a proof of $A \vee B$, will output a length $p(l)$ proof of one or the other.

Statements in complexity theory can be reformulated as families of propositional statements. In order to get rid of all the quantifiers used in a standard mathematical statement, we will allow the propositional equivalent to be exponentially longer than the original. For example, the statement that satisfiability requires circuits of size $\Omega(n^{(\log n)})$ can be handled as follows. S will be a propositional statement of size $2^{O(n)}$ asserting that there is no circuit of size $n^{\log n}$ that can decide satisfiability for all formulas of length n. S has two easy building blocks. (1) A formula SAT which tells us which formulas of length n are satisfiable. SAT is trivial to construct because we allow it to be of length 2^n. (2) A formula CVAL that takes an $O(n^{\log n})$ bit index C into one of the circuits of size $n^{\log n}$ as well as an input X to the circuit and returns true iff $C(X) = 1$. CVAL can be easily constructed with size $O(n^{2\log n})$. Let X_i for $1 \leq i \leq 2^n$ be an enumeration of n-bit strings. $C(X_i)$ will be an abbreviation for CVAL(C, X_i). The free variables of S will be $O(n^{\log n})$ bits which should be thought of as indexing a circuit C. $S(C) \equiv (C(X_1) \neq SAT(X_1)) \vee (C(X_2) \neq SAT(X_2)) \vee \cdots \vee (C(X_n) \neq SAT(X_n))$. S has size $O(2^{2n})$. Let $\langle S_m \rangle$ be the family of propositions where S_m ($m = O(2^{2n})$) asserts that SAT on inputs of length n does not have $n^{\log n}$ size circuits. The brute force way of proving S_m would involve enumerating all the $2^{n^{\log n}}/n^{\log n} = o(2^{kn})$ circuits of size $n^{\log n}$. The burning question is whether or not standard boolean algebra allows for a polynomial-size proof of $\langle S_m \rangle$. A negative result would imply that S_2^1 can't prove SAT$\notin$P/poly.

Theorem [Razborov [8]]: If a reasonable[1] propositional proof system with CIP proves $\langle S_m \rangle$ with proofs of polynomial-size, then there is a combinatorial property that is $\tilde{P}$-natural against $P/poly$.

Corollary: If 2^{n^ε}-hard functions exist, then $\langle S_m \rangle$ does not have polynomial-size proofs in any reasonable propositional proofs systems with CIP.

An easy corollary of Razborov's proof shows:

Theorem 1. *If a reasonable propositional proof system with EIP proves $\langle S_m \rangle$ with proofs of polynomial-size, then there is a combinatorial property that is $N\tilde{P}/qpoly$-natural against $P/poly$.*

Corollary: If Conjecture 1 is true, then $\langle S_m \rangle$ does not have polynomial-size proofs in any reasonable propositional proofs systems with EIP.

Thus, the project of extending the study of $\tilde{P}/qpoly$ natural proofs to $N\tilde{P}/qpoly$-natural corresponds to extending the study of propositional systems with CIP to those with EIP.

5 Cryptography against nondeterministic adversaries.

5.1 The super-bit conjecture.

At first glance, it would seem impossible to generalize the notion of a pseudorandom generator to work against nondeterministic adversaries. The reason is simple: a nondeterministic adversary can guess the seed of the generator and verify that the guess is correct.

In this section, we will show how to change the standard definition of pseudorandom generator to handle the case of nondeterministic adversaries. Furthermore, we show that the known generator constructions still work according to this stronger definition.

---

[1] Reasonable requires a few simple properties common to all studied propositional proof systems.

We start by reviewing the standard definition of pseudorandomness in the non-uniform circuit model.

Definition 2. (Standard Hardness and Pseudorandom Generators): Let $g_n : \{0,1\}^n \rightarrow \{0,1\}^{l(n)}$ be a family in $P/poly$ where $l(n) > n$. The hardness $H(g_n)$ of pseudorandom generator g_n is the minimal S for which there exists a circuit C of size $\leq S$ such that

$$|Pr_{z \in \{0,1\}^{l(n)}}[C(z) = 1] - Pr_{x \in \{0,1\}^n}[C(g_n(x)) = 1]| \geq \frac{1}{S}$$

Here is the standard hardness conjecture that was shown to imply that $P/poly$-natural properties against $P/poly$ do not exist [10].

Conjecture 2 *(Standard Conjecture)*: *There is a pseudorandom generator with hardness 2^{n^ϵ}, for some $\epsilon > 0$.*

Notice that the use of the absolute value does not really matter. This because (especially in the non-uniform model) it is easy to make an adversary circuit C' that determines for a given n if the difference is positive or negative and flips C's output accordingly. Thus, we can not only drop the absolute value, but we can also *order the two probabilities in the difference either of two ways.*

To obtain a stronger definition that uses nondeterministic adversaries, we simply drop the absolute value! Unlike the deterministic case, the order of the two probabilities is *crucial.*

Definition 3. (Non-deterministic Hardness) Let $g_n : \{0,1\}^n \rightarrow \{0,1\}^{l(n)}$ be a family in $P/poly$ where $l(n) > n$. The hardness $H(g_n)$ of pseudorandom generator g_n is the minimal S for which there exists a non-deterministic circuit C of size $\leq S$ such that

$$Pr_{z \in \{0,1\}^{l(n)}}[C(z) = 1] - Pr_{x \in \{0,1\}^n}[C(g_n(x)) = 1] \geq \frac{1}{S}$$

If we reversed the order of the probabilities or kept the absolute value, a simple adversary would *always* break the generator: guess the seed x and verify that it is consistent with the observed string.

By placing $Pr_{z \in \{0,1\}^{l(n)}}[c(z) = 1]$ first, we force a nondeterministic circuit to prove that an observed string is probably random, as opposed to proving that the observed string is probably the output of the generator. Non-deterministic circuit classes are not likely to be closed under complement. This asymmetry is what makes the order of the probabilities so important.

Equivalently, we could formulate it in such a way that the two probabilities range over disjoint sets:

$$Pr_{z \in \{0,1\}^{l(n)}, z \notin range_g}[C(z) = 1] - Pr_{x \in \{0,1\}^n}[C(g_n(x)) = 1] \geq \frac{1}{S}$$

In other words, there is a function $g \in P/poly$ such that C cannot be true on a significantly larger fraction of non-range elements than the fraction of range elements it is true on.

After consideration, we make the follow conjecture. It could also be called the *non-deterministic hardness* conjecture.

Conjecture 3 *(Super-bit Conjecture)* *There exists a* $g : \{0,1\}^n \to \{0,1\}^{n+1}$ *in P/poly with non-deterministic hardness* 2^{n^ϵ}, *for some* $\epsilon > 0$.

We call it the super-bit conjecture because it says that we can produce a bit of pseudorandomness (n bits in/ $n + 1$ bits out) that can fool a super powerful adversary.

We note that a random function oracle $g_n : \{0,1\}^n \to \{0,1\}^{n+1}$ can easily be seen to satisfy the definition of a super-bit. In other words, the conjecture is true relative to a random oracle.

5.2 A super-bit based on subset sum.

We propose that a previously studied generator[4], the subset sum generator, is already the source of a secure super-bit generator. (We are grateful to Moni Naor for this suggestion.)

Let g be a function taking m $m + 1$-bit numbers, $a_1, a_2, \ldots, a_m$ and m 1-bit numbers, $b_1, b_2, \ldots, b_m$.

$$g(a_1, a_2, \ldots, a_n, b_1, b_2, \ldots, b_n) = a_1 b_1 + a_2 b_2 + \cdots + a_m b_m \quad \text{MOD } 2^{m+1}$$

Notice that g is a function that takes $m(m + 1) + m$ bits of input and returns $m(m + 1) + m + 1$ bits of output.

We conjecture that this function has the super-bit requirement. Intuitively, it is very hard to prove that a given $n + 1$-bit sequence is not in the range of g. The length of the shortest proof of non-solvability of subset sum was considered in Furst and Kannan [2]. They were not able to find shorter-than-obvious proofs for subset sum problems on the instance sizes used in the definition of g.

Conjecture 4 *(Subset Sum Conjecture)* *Let* $g : \{0,1\}^n \to \{0,1\}^{n+1}$ *be the subset sum function above.* g *has non-deterministic hardness* 2^{n^ϵ}, *for some* $\epsilon > 0$.

5.3 Stretching a super-bit.

In the standard theory of pseudorandomness, a single bit of pseudorandomness can be stretched to many bits [1, 11]. In fact, Goldreich, Goldwasser, and Micali[3] showed that it is possible to construct pseudorandom function generators based on any pseudorandom bit generator. In this section, we argue that, using standard constructions, a single super-bit can be stretched to many super-bits, and even used to create pseudorandom function generators with non-deterministic hardness. The proofs are omitted in this abstract. It suffices to peruse the standard proofs [7] and to notice that they never flip the order of the terms in the difference.

Let $g : \{0,1\}^n \to \{0,1\}^{n+1}$ be a function. We stretch g using the standard construction [7]. Let $first(x)$ be the leftmost bit of x, and $rest(x)$ be the remaining $|x| - 1$ rightmost bits. $g^0(x) = x, g^1(x) = g(x)$ and, for all $i \geq 1$, $g^{i+1}(x) = first(g^i(x)); g^i(rest(g(x)))$.

Theorem 4. *If g is a pseudorandom generator with non-deterministic hardness S, then $g^{l(n)}$ is a pseudorandom generator with non-deterministic hardness $S/l(n)$.*

We are now in a position to make a pseudorandom function generator. We know that assuming a super-bit generator, we can make a generator $g : \{0,1\}^n \to \{0,1\}^{2n}$ that doubles the length of its input. If x is an even length string, let $left(x)$ be the left half of x, and $right(x)$ be the right half of x. Let $g^0(x) = x$, $g^1(x) = g(x)$ and, for all $i \geq 1$, $g^{i+1}(x) = g^i(left(g(x)))$; $g^i(right(g(x)))$.

Theorem 5. *If g is a pseudorandom generator with non-deterministic hardness S, then $g^{l(n)}$ is a pseudorandom generator with non-deterministic hardness $S/(2^{l(n)})$.*

Furthermore, $g^{l(n)}$ gives us the basis for a pseudorandom function generator. Define $node_x(emptystring) = x$. For a non-empty string y of length less than or equal to $l(n)$, define $node_x(y) = left(g(node_x(rest(y))))$ if $first(y) = 0$, and $right(g(node_x(rest(y))))$ if $first(y) = 1$. Define $f_x(y) = first(node_x(y))$. Notice that, as long as x and y have polynomially related lengths, there is a polynomial-size circuit family to compute $f_x(y)$, given x and y. Define $F(x)$ to be the bit sequence that gives the truth table of f_x.

Corollary to Theorem 5: If g is a pseudorandom generator with non-deterministic hardness S, then F is a pseudorandom generator with output length $2^{l(n)}$ and non-deterministic hardness at least $S/2^{l(n)}$.

This is a trivial corollary, since $F(x)$ presents strictly less information to an adversary than does $g^{l(n)}(x)$. Therefore, we can view $f_x(y)$ as a pseudorandom function family. For each x, f_x can be computed by a polynomial-size circuit family, and for a random x, $F(x)$ looks random to any non-deterministic adversary.

5.4 $N\tilde{P}/qpoly$-natural properties and the super-bit conjecture.

Assume that we can generate a super-bit, then by Theorem 4 we know that we have a pseudorandom generator $g : \{0,1\}^r \to \{0,1\}^{r+1}$ with non-deterministic hardness 2^{n^ϵ}, for some fixed $\epsilon > 0$. Assume that Conjecture 1 is false, i.e., there exists an $N\tilde{P}/qpoly$-natural property C_n that is useful against $P/poly$. We will obtain a contradiction by following reasoning similar to [10] and [8].

Without loss of generality, we can assume that C_n already satisfies the largeness condition (as opposed to some subset C_n^*). Define $N = 2^n$. We can choose an integer k such that for sufficiently large n, the bound on the size of the circuit that computes C_n is less than $N^{\log^{k-2}(N)} = 2^{n^{k-1}}$, and that $|C_n| > 2^{-n^{k-1}} \cdot |F_n| = N^{-\log^{k-2}(N)} \cdot 2^N$.

Let $m = n^{k/\epsilon}$. Choosing $|x| = m$ and $|y| = m^{\epsilon/k} = n$, we can use the construction from Theorem 5 to obtain a pseudorandom function family f_x with non-deterministic hardness $2^{m^\epsilon}/2^n = 2^{n^{k-1}}$ such that $|F(x)| = N$.

Since each function f_x is in $P/poly$, we know that for all x $C_n(F(x)) = 0$. By assumption, $|C_n| > 2^{-n^{k-1}} \cdot |F_n|$. Thus,

$$|Pr_{z \in \{0,1\}^N}[C_n(z) = 1] - Pr_{x \in \{0,1\}^m}[C_n(F(x)) = 1]| > \frac{1}{2^{n^{k-1}}}$$

The circuit C_n that performs this statistical test has size less than $2^{n^{k-1}}$. This contradicts the non-deterministic hardness of f_x.

We conclude:

Theorem 6. *If super-bits exist (Conjecture 3), then Conjecture 1 (no $N\tilde{P}/qpoly$-properties) is true.*

6   The Demi-bit conjecture.

The notion of a super-bit has the advantage of being the generalization of a pseudorandom bit. Unfortunately, Conjecture 3 will be less than intuitive to some people. We seek here to develop a more intuitive conjecture that might also be sufficient to rule out $N\tilde{P}/qpoly$-natural properties. In fact, we are unable to show that demi-bits can be stretched, as can super-bits. Whether useful in this context or not, demi-bits are of independent interest.

Definition 7. (Demi-Hardness) Let $g_n : \{0,1\}^n \to \{0,1\}^{l(n)}$ be a family in $P/poly$ where $l(n) > n$. The demi-hardness $H(g_n)$ of pseudorandom generator g_n is the minimal S for which there exists a non-deterministic circuit C of size $\leq S$ such that

$$Pr_{z \in \{0,1\}^{l(n)}}[C(z) = 1] \geq \frac{1}{S}$$

and

$$Pr_{x \in \{0,1\}^n}[C(g_n(x)) = 1] = 0$$

Conjecture 5 *(demi-bit Conjecture)* *There exists a $b : \{0,1\}^n \to \{0,1\}^{n+1}$ in $P/poly$ with demi-hardness 2^{n^ϵ}, for some $\epsilon > 0$.*

This is more intuitive. Simply, it says that there is function such that most non-range elements have no short certificate that prove that they are not in the range of b. This is similar to formalizing the notion of a $coNP$ predicate that is average case hard for NP.

Open Problem 1 *Does the demi-bit conjecture imply the super-bit conjecture?*

Open Problem 2 *If you have one demi-bit, can you stretch it to 2 demi-bits?*

Open Problem 3 *If you have one demi-bit, can you build a pseudorandom function generator with demi-hardness 2^{n^ϵ}?*

A positive answer to this last problem would solve the next one:

Open Problem 4 *Can you prove that the demi-bit conjecture rules out the existence of $N\tilde{P}/qpoly$-natural properties?*

7 Other Open Problems

Open Problem 5 *On what weaker assumption can one prove the existence of a demi-bit?*

Open Problem 6 *On what weaker assumption can one prove the existence of a super-bit?*

8 Acknowledgements

We would like to thank Moni Naor, Avi Wigderson, Sasha Razborov, Russell Impagliazzo, Oded Goldreich, and Jiří Sgall for useful conversations.

References

1. M. Blum and S. Micali. How to generate cryptographically strong sequences of pseudo-random bits. *SIAM Journal on Computing*, 13:850–864, 1984.
2. M. Furst and R. Kannan. Succinct certificates for almost all subset sum problems. *Siam Journal on Computing*, vol 18, 1989, pp. 550-558.
3. O. Goldreich, S. Goldwasser, and S. Micali. How to construct random functions. *Journal of the ACM*, 33(4):792–807, 1986.
4. R. Impagliazzo and M. Naor. Efficient Cryptographic Schemes Provably as Secure as Subset Sum. *Journal of Cryptology*, 9(4):199-216, 1996.
5. J. Krajíček. *Bounded Arithmetic, Propositional Logic, and Complexity Theory*, Cambridge University Press, 1995.
6. J. Krajíček. Interpolation theorems, lower bounds for proof systems and independence results for bounded arithmetic. Submitted to *Journal of Symbolic Logic*, 1994.
7. Michael Luby, *Pseudorandomness and Cryptographic Applications*, Princeton University Press (1996).
8. A. Razborov. Unprovability of lower bounds on circuit size in certain fragments of Bounded Arithmetic. *(Izvestiya of the RAN)*, 59(1):201–222, 1995.
9. A. Razborov. On provably disjoint **NP**-pairs. Technical Report RS-94-36, Basic Research in Computer Science Center, Aarhus, Denmark, 1994.
10. A. Razborov and S. Rudich. Natural Proofs. *26th Annual Symposium on Theory of Computing* (1994). To appear in special issue of JCSS.
11. A. Yao. Theory and Applications of Trapdoor Functions. *Proc. 23rd IEEE Symp. on Foundations of Computer Science*, Chicago, IL (Nov. 1982), 80–91.

Sample Spaces with Small Bias on Neighborhoods and Error-Correcting Communication Protocols[*]

Michael Saks[1], Shiyu Zhou[2]

[1] Department of Mathematics, Rutgers University, New Brunswick, NJ 08854, USA.
E-mail: *saks@math.rutgers.edu*
[2] Bell Laboratories, 700 Mountain Ave., Murray Hill, NJ 07974, USA. E-mail:
shiyu@research.bell-labs.com

Abstract. We give a deterministic algorithm which, on input an integer n, collection $\mathcal{F}$ of subsets of $\{1, 2, \ldots, n\}$ and $\epsilon \in (0, 1)$ runs in time polynomial in $n|\mathcal{F}|/\epsilon$ and produces a ± 1-matrix M with n columns and $m = O(\log |\mathcal{F}|/\epsilon^2)$ rows with the following property: for any subset $F \in \mathcal{F}$, the fraction of 1's in the n-vector obtained by coordinate-wise multiplication of the column vectors indexed by F is between $(1 - \epsilon)/2$ and $(1 + \epsilon)/2$. In the case that $\mathcal{F}$ is the set of all subsets of size at most k, k constant, this gives a polynomial time construction for a k-wise ϵ-biased sample space of size $O(\log n/\epsilon^2)$, as compared to the best previous constructions of [NN90] and [AGHP91] which were, respectively, $O(\log n/\epsilon^4)$ and $O((\log n)^2/\epsilon^2)$. The number of rows in the construction matches the upper bound given by the probabilistic existence argument. Such constructions are of interest for derandomizing algorithms.
As an application, we present a family of essentially optimal deterministic communication protocols for the problem of checking the consistency of two files.

1 Introduction

Suppose A is a randomized algorithm for a given problem P and that on a given input x, A requires n random bits. Then A can be derandomized by "brute force" by running the algorithm on x with each of the 2^n possible random strings and combining the results in an appropriate way. A general approach to improving this brute force derandomization is to identify a small subset $S_n(A)$ of n-bit strings such that running A with each of the strings in the subset suffices to answer $P(x)$. Underlying this approach is the observation that for a typical randomized algorithm, the proof of correctness does not require that the sequence $(X_1, X_2, \ldots, X_n)$ of random bits come from the uniform distribution over all strings, rather it is sufficient that the sequence be selected from any probability distribution that satisfies some specified "random-like" property. The specific property required depends on the algorithm and its analysis; one example of such a property that commonly arises is *unbiased k-wise independence* for some $k < n$,

---

[*] This work was supported in part by NSF under grant CCR-9215293 and by DIMACS.

which means that for any subset of k of the bits, the distribution restricted to those bits is uniform over $\{-1, 1\}^k$ (for technical convenience we will use $\{-1, 1\}$ as our set of Boolean values in most of this paper). Given any distribution D that satisfies the appropriate property for a given algorithm, we can take $S_n(A)$ to be $support(D)$, the set of strings assigned nonzero probability by D.

This gives the following two step approach to derandomizing an algorithm: (1) isolate a property of probability distributions that suffices for the analysis of the algorithm, and (2) give an efficient algorithm for constructing a distribution that has these properties and also has small size (where by "size" we mean the size of the support). This general approach has been used successfully to derandomize a variety of randomized algorithms, see e.g. [KW84, Lub85, ABI86, BR89, MNN89, NN90, Alo91, Sch92, KM93, KK94], and motivates the problem of constructing distributions satisfying certain specified constraints and having small support. An efficient construction of an unbiased k-wise independent distribution having support size $n^{\lfloor k/2 \rfloor}$ was presented in [ABI86] (see also [Lub85]); the size essentially matches the lower bound given in [CGHFRS85] (see also [ABI86, KM94]). More generally, Schulman [Sch92] considered the problem of constructing probability distributions uniform over an arbitrary family $\mathcal{F}$ of subsets (called neighborhoods) of $\{1, 2, \ldots, n\}$. A distribution is uniform over $\mathcal{F}$ if for any $F \in \mathcal{F}$ the distribution restricted to the random variables indexed by F is uniform on $\{-1, 1\}^{|F|}$. In the case that $\mathcal{F}$ is the set of all subsets of size k, this is the same as unbiased k-wise independence. In Schulman's construction the size of the support of the distribution is $O(d2^l)$, where l is the maximum size of any neighborhood in the family and d is the maximum number of neighborhoods containing any element.

Another property of interest in this context is ϵ-bias. For a given distribution D on n-bit strings, the *bias* ([Vaz86]) of a subset $J \subseteq \{1, 2, \ldots, n\}$ relative to D is equal to $|E[\prod_{i \in J} X_i]|$ where $E[\cdot]$ denotes expectation. Naor and Naor [NN90] introduced the notion of an ϵ-biased (resp. k-wise ϵ-biased) probability distribution, in which the bias of every subset (resp. every subset of size at most k) of $\{1, 2, \ldots, n\}$ is at most ϵ. They presented an efficient construction of an ϵ-biased distribution with support size $O(n/\epsilon^4)$. Alon et al. in [AGHP91] gave three simple constructions of ϵ-biased distributions of size $O(n^2/\epsilon^2)$. These constructions also yield k-wise ϵ-biased distributions of size $O(k \log n/\epsilon^4)$ and $O(k^2 \log^2 n/\epsilon^2)$, respectively. In [EGLNV92], constructions of distributions with small bias for general independent distributions were studied.

Koller and Megiddo in [KM93] introduced a more general framework of constructing distributions that satisfy a set of linear probability constraints, generalizing the construction approach used in [Sch92]. This method was further extended by Karger and Koller in [KK94] where they showed how to deal with the more general expectation constraints. In the latter work, an NC algorithm for constructing distributions that approximate general independent distributions with relative error was shown.

In this paper, we extend the approaches of Schulman and Naor-Naor by considering the problem of constructing distributions that are ϵ-biased on a given family $\mathcal{F}$. For a given family $\mathcal{F}$ of subsets of $\{1, 2, \ldots, n\}$, we construct a probability distribution with small support such that the bias of any $F \in \mathcal{F}$ is at most ϵ.

Denote the maximum cardinality of any element of $\mathcal{F}$ by $\Upsilon(\mathcal{F})$. The two best previously known constructions for this problem are of sizes $O(\Upsilon(\mathcal{F}) \log n / \epsilon^4)$ [NN90] and $O(\Upsilon(\mathcal{F})^2 \log^2 n / \epsilon^2)$ [AGHP91], respectively, which are obtained by constructing $\Upsilon(\mathcal{F})$-wise ϵ-biased sample spaces. We present a new construction whose size is $O(\log |\mathcal{F}| / \epsilon^2)$. The running time of our construction is polynomial in n, ϵ^{-1} and $|\mathcal{F}|$. Since $|\mathcal{F}| \leq n^{\Upsilon(\mathcal{F})}$ always holds, the size of our construction can never exceed the order of the sizes of the previous constructions. In the case that $\mathcal{F}$ is of size polynomial in n, our construction runs in polynomial time and has size $O(\log n / \epsilon^2)$ which is independent of $\Upsilon(\mathcal{F})$. In particular, for fixed k we obtain a k-wise ϵ-biased sample space of size $O(\log n / \epsilon^2)$ which is the best known construction.

The idea of the construction is as follows. It is not difficult to show that if we select $O(\log(|\mathcal{F}|)/\epsilon^2)$ points uniformly at random from $\{0, 1\}^n$ then, with high probability, the empirical distribution induced by this sample is ϵ-biased on $\mathcal{F}$. This observation proves the *existence* of a distribution that is ϵ-biased on $\mathcal{F}$ having support size $O(\log(|\mathcal{F}|)/\epsilon^2)$, and also gives a randomized algorithm for finding such a distribution. Thus, the problem of deterministically constructing such a distribution can itself be viewed as the problem of derandomizing an algorithm. For this derandomization we will use the method of conditional probabilities ([ES73, Spe87, Rag88]). However, applying the method directly to the above randomized algorithm results in an exponential time algorithm. Instead we show that the above randomized algorithm has high probability of producing a distribution with the required property provided that the $O(\log |\mathcal{F}| / \epsilon^2)$ points are selected according to any distribution D that is $\frac{\epsilon}{2}$-biased. By using such a distribution D having polynomial support size (as noted above, [NN90] and [AGHP91] gave explicit constructions of such distributions), we can then apply the method of conditional probabilities to get a polynomial time construction of the distribution we seek.

In the second part of this paper, we use our construction to obtain error-correcting communication protocols for comparing two files at separated locations. Often in a distributed system, several physical copies of a single file are maintained at distinct locations. These copies are supposed to be identical, but discrepencies can occur, for example, because of undetected hardware failure or bugs in the update protocols. In this situation, the system must perform periodic consistency checks to ensure that the various copies are the same.

We consider the case that there are two copies of the file, and that communication between them is expensive compared to the local computation. In designing a protocol to detect or correct inconsistencies, we seek to minimize the communication. It is a well known fact in the theory of communication complexity that if we make no assumptions about the state of the two files, then in worst case no protocol is better than transmitting the entire file at one location to the other location. This seems overly pessimistic since, typically, one expects that the inconsistencies between the two databases will be very sparse. For example, it might be safe to assume that the fraction of bits on which they differ is at most $1/10000$ or the file may be organized into some number of blocks, and we can assume that at most $1/500$ of the blocks are corrupted. More generally, depending on the file layout and empirical error patterns, we can determine a family of sets of bit positions (difference patterns) such that, in (almost) all the cases, the set of positions where the inconsistencies occur is a member in the family. We may then look at the problem of designing a protocol that is capable of correcting discrepancies under the assumption that they occur in one of the predicted difference patterns. (Since there is a small chance that the discrepancy falls outside the predicted difference pattern, we would still need occasionally to perform more thorough, e.g., exhaustive, comparisons of the two databases.)

Abstractly, we have the following communication problem. Let $\mathcal{F}$ be a fixed family of subsets of $\{1, 2, \ldots, n\}$ which includes $\emptyset$. Two parties A and B with limited computation power (polynomial time machines) are given an n-bit string x and an n-bit string y respectively such that the set of positions on which x and y differ is a member in $\mathcal{F}$. Knowing $\mathcal{F}$, A and B want to determine whether or not x and y are equal (error-detection), and in the case they are not, A and B want to tell the *exact* positions in which x and y differ (error-correction). The goal is to minimize the communication.

The error detection case of this problem is a natural variant of the *equality* problem in the study of communication complexity (see e.g. [Yao79]), in which A and B have unlimited computation power and try to tell whether or not x and y are the same (using the above language, $\mathcal{F}$ contains all subsets of $\{1, 2, \ldots, n\}$).

Let $\mathcal{F}^*$ denote the set of all symmetric differences of sets in $\mathcal{F}$ (so $|\mathcal{F}^*| \leq |\mathcal{F}|^2$. We show that a distribution D that is $1/2$-biased with respect to $\mathcal{F}^*$ can be used as the basis of both an error detection and an error correction protocol for this problem, whose complexity is $O(|support(D)|)$.

Section 2 presents some basic notation and definitions. In Section 3 we present our construction of sample spaces with small bias on neighborhoods. Finally in Section 4 we show to use our construction to design error-correcting communication protocols and also to improve on the known constructions for *universal sets*.

2 Preliminaries

2.1 Basic Notation

For integer $n \geq k \geq 0$, $[n]$ denotes the set of integers $\{1, 2, \ldots, n\}$ and $\binom{[n]}{k}$ denotes the family of subsets of $[n]$ of size k. If $\mathcal{F}$ is a family of sets, $\Upsilon(\mathcal{F})$ denotes the maximum size of any set in the family. $\Lambda(\mathcal{F})$ is defined to be $\{J : J \subseteq F$ for some $F \in \mathcal{F}\}$, the family of all subsets of sets in $\mathcal{F}$. For example, $\Lambda(\binom{[n]}{k})$ consists of all subsets of $[n]$ having size at most k. In the case that $\mathcal{F}$ is a singleton set $\{J\}$, we sometimes write $\Lambda(J)$ instead of $\Lambda(\{J\})$. For example, $\Lambda([n])$ is the family of all subsets of $[n]$.

For a vector $v = (v_1, v_2, \ldots, v_n)$, we define v_+ to be the sum of its entries. If v has n entries and $J \subseteq [n]$ we define $v_J = \prod_{j \in J} v_j$.

If M is a $m \times n$ matrix, we denote by $M_{i,*}$ the ith row vector and $M_{*,j}$ the jth column vector. The above notation for vectors is extended as follows: $M_{+,j}$ is the sum of the entries in the jth column, for $J \subseteq [n]$, $M_{*,J}$ is the n-vector whose i^{th} entry is $M_{i,J} := \prod_{j \in J} M_{i,j}$, and $M_{+,J}$ is the sum of the entries in $M_{*,J}$.

A probability distribution D on a finite set S is a function assigning to each $s \in S$ a nonnegative probability $D(s)$, with $\sum_{s \in S} D(s) = 1$. We write $E_D[\cdot]$ and $Pr_D[\cdot]$ for expectation and probability with respect to D, and omit the D if it is clear from context. We define $support(D) = \{s \in S : D(s) > 0\}$. If m is a positive integer D^m, the m-fold power of D, is the distribution on S^m corresponding to selecting a sequence $s_1, s_2, \ldots, s_m$ where each s_i is chosen independently according to D.

We associate to each $m \times n$ matrix M a probability distribution D_M which selects a row of M uniformly at random; if v is an n-vector then $D_M(v)$ is proportional to the number of rows of M that are equal to v. A common abuse of terminology is to refer to the multiset of rows of M as the *sample space* for the distribution D_M.

Throughout this paper, most of the vectors and matrices we encounter will have entries in the set $\{-1, +1\}$. we call such a matrix (resp. vector) a $\pm$-matrix (resp. $\pm$-vector).

2.2 Sample Spaces with Small Bias

For a distribution D on $\{-1, 1\}^n$ and a subset $F \subseteq [n]$, the *bias of F relative to D* is defined to be $|E_D[\prod_{i \in F} X_i]|$ (which equals $|Pr_D[\prod_{i \in F} = 1] - Pr_D[\prod_{i \in F} = -1]|$).

Definition 2.1 *Let D be probability distribution on $\{-1, 1\}^n$. For $F \subseteq [n]$, D is ϵ-biased on F if the bias of F relative to D is at most ϵ. For a collection $\mathcal{F}$ of subsets of $[n]$, D is ϵ-biased on $\mathcal{F}$ if it is ϵ-biased on each $F \in \mathcal{F}$. D is said to be k-wise ϵ-biased if it is ϵ-biased on $\Lambda(\binom{[n]}{k})$, and everywhere ϵ-biased, or simply ϵ-biased, if it is ϵ-biased on $\Lambda([n])$.*

We extend these definition to $\pm$-matrices M: if M has n columns we say that M is ϵ-biased (on the set F or the collection $\mathcal{F}$) if the distribution D_M is. By the definition of D_M we have:

Proposition 2.1 *Let M be a $m \times n$ $\pm$-matix and $F \subseteq [n]$. The following are equivalent:*

1. *M is ϵ-biased on F.*
2. *The fraction p_F of 1's in the (column) vector $M_{*,F}$ is between $(1 - \epsilon)/2$ and $(1 + \epsilon)/2$.*
3. *$|M_{+,F}| \leq \epsilon m$.*

We will need the next result due from [NN90] (see also [AGHP91]):

Theorem 2.1 *Let n be a positive integer and $0 < \epsilon \leq 1$. There is a deterministic algorithm running in time $\text{poly}(\frac{n}{\epsilon})$ that produces a $m' \times n$ $\pm$-matrix M' with $m' = \text{poly}(\frac{\log n}{\epsilon})$ that is everywhere ϵ-biased.*

2.3 The Method of Conditional Probabilities

The method of conditional probabilities, introduced by Erdös and Selfridge[ES73] and further developed by Spencer[Spe87] and Raghavan [Rag88], is a powerful tool for derandomization. Here we review one framework for this method.

Let S be a set, m be a positive integer and Z be a real valued function on S^m. We wish to find a point $v \in S^m$ such that $Z(v) < c$, where c is some given constant. Suppose that we have a probability distribution D on S^m such that $E[Z] < c$. Then there must be at least one point $v \in S^m$ for which $Z(v) < c$. The method of conditional probabilities gives a deterministic search procedure for finding such a point, which is efficient in many situations.

Associated to the distribution D is the sequence $X_1, X_2, \ldots, X_m$ of S-valued random variables where X_i denotes the ith coordinate of the point selected according to D. For $1 \leq i \leq m$ and $v_1, \ldots, v_i \in S$, write $E[Z|v_1, \ldots, v_i]$ to denote the conditional expectation of Z given $X_1 = v_1, \ldots, X_i = v_i$. We then observe that for any $i \in [m]$ and $(v_1, \ldots, v_{i-1}) \in S^{i-1}$, $E[Z|v_1, \ldots, v_{i-1}] \geq \min_{s \in S} E[Z|v_1, v_2, \ldots, v_{i-1}, s]$. This follows since the term on the left can be written as a convex combination of the terms that appear inside the minimum on the right. This suggests the following greedy approach for choosing $v_1, v_2, \ldots, v_m$: for each $i \in [m]$, select v_i (having already chosen $v_1, v_2, \ldots, v_{i-1}$) to be the value of s minimizing $E[Z|v_1, v_2, \ldots, v_{i-1}, s]$. The above inequality guarantees that this produces a point v such that $Z(v) < c$.

The complexity of this algorithm depends on the cost of computing the conditional probabilities that arise. If for each $1 \leq i \leq m$ and $v_1, \ldots, v_i \in S$, the conditional expectation $E[Z|v_1, \ldots, v_i]$ is computable in time T_Z, then the selection of v_i given $v_1, v_2, \ldots, v_{i-1}$ requires at most $|S|$ such expectations, and so the selection of s takes time $O(mT_Z|S|)$. The time can be reduced by observing that in selecting s_i we only need to check those s for which $Pr[X_i = s|X_1 = v_1, \ldots, X_{i-1} = v_{i-1}]$ is nonzero.

Remark 2.1 *In specific applications of this method, the function Z may not be computable with exact precision, for instance it may involve transcendental functions such as e^x. In such cases, we may only be able to estimate each conditional expectation to within some error μ. Starting with $E[Z] < c$, we can only guarantee that the point constructed satisfies $Z(v) < c + m\mu$. Thus, to ensure that we get a point with $Z(v) < c$ we need to start with $E[Z] < c'$ for some $c' < c$ and do the calculations with enough precision that the error in each iteration is at most $(c - c')/m$.*

In many circumstances, the conditional expectations of Z are neither efficiently computable nor approximable. One effective technique to overcome this problem is the method of *pessimistic estimators*, introduced by Raghavan [Rag88]. The idea (as formulated by Fundia [Fun94]) is simple: in applying the method of conditional probabilities to find a v such that $Z(v) < c$ we can replace Z by any function W on S^m that satisfies:

(A) $Z(v) \leq W(v)$ for all $v \in S^n$,
(B) $E[W] < c$,

since these two conditions guarantee that the method applied with W produces a point v such that $Z(v) < c$. The function W is called a *pessimistic estimator* for Z. If we can find any such W for which the associated conditional expectations are efficiently computable, we can obtain an efficient deterministic construction for the desired v.

3   Sample Spaces with Small Bias on Neighborhoods

In this section, we present a deterministic algorithm that, given a set $\mathcal{F}$ of subsets of $[n]$ and a number $\epsilon \in (0, 1)$ runs in time polynomial in $n|\mathcal{F}|/\epsilon$ and produces a $\pm$-matrix with $O(\log |\mathcal{F}|/\epsilon^2)$ rows that is ϵ-biased on $\mathcal{F}$.

The idea behind our construction is as follows. A standard probabilistic (counting) argument shows that if we select a random $\pm$-matrix M with $m = O(\log |\mathcal{F}|/\epsilon^2)$ rows and n columns, then with high probability it is ϵ-biased on $\mathcal{F}$. The random process for selecting M is equivalent to the process that proceeds row by row, choosing each row independently according to the uniform distribution U on $\{-1, 1\}^n$, so that M is chosen according to the m-fold power distribution U^m. We would like to apply the method of conditional probabilities to build such a matrix M deterministically by selecting its rows one by one. We must overcome several hurdles to do this. First we need a function Z on the set of $m \times n$ $\pm$-matrices and a constant c with the following two properties: (i) for any matrix M, $Z(M) \leq c$ implies that M is ϵ-biased on $\mathcal{F}$, and (ii) $E[Z] \leq c$. A natural choice for Z is to take $Z(M)$ to be the number of subsets $F \in \mathcal{F}$ such that M is not ϵ-biased on F; $Z(M) < 1$ if and only if M is ϵ-biased on $\mathcal{F}$, and the probabilistic existence argument shows that $E[Z] < 1$. Thus the method of conditional probabilities can be applied. However the complexity of the method $O(mT_Z|S|)$ is very high for two reasons: computing the conditional probabilities for Z seems to be hard, and $|S| = 2^n$. The first problem is handled by constructing a pessimistic estimator W such that T_W is polynomial in $n|\mathcal{F}|/\epsilon$. The second problem is handled by returning to the original existence argument and noting that it can be generalized so that essentially the same conclusion holds if the rows of M are selected independently from any everywhere δ-biased distribution D' with $\delta < \epsilon$. Taking $\delta = \epsilon/2$, by Theorem 2.1 we can deterministically construct a matrix $m' \times n$ matrix M' that is everywhere δ-biased and $m' = poly(\frac{n}{\epsilon})$. The support of the distribution $D' = D_{M'}$ has size equal to the number of distinct rows of M' and hence $|S|$ is reduced to polynomial in $\frac{n}{\epsilon}$. So we obtain our algorithm by applying the method of conditional probabilities with the distribution D' and the pessimistic estimator W.

3.1 The Existence Proof

Lemma 3.1 *Let D' be a δ-biased distribution over $\{-1,1\}^n$ and let $\mathcal{F}$ be a family of subsets of $[n]$. Then for any $\epsilon > \delta$, the random matrix M obtained by selecting $\frac{2\ln(4|\mathcal{F}|)}{(\epsilon-\delta)^2}$ rows independently according to the distribution D' has the following property: the expected number of sets $F \in \mathcal{F}$ such that M is not ϵ-biased on F is less than $1/2$. Hence the probability that M is ϵ-biased on $\mathcal{F}$ is greater than $1/2$.*

Proof: Let $m = \frac{2\ln(4|\mathcal{F}|)}{(\epsilon-\delta)^2}$ and let M be the $m \times n$ matrix produced by the random process in the hypothesis. By Proposition 2.1 M is ϵ-biased on F if and only if $|M_{+,F}| \leq \epsilon m$. For $F \in \mathcal{F}$ let Z_F be the random variable which is 1 if $|M_{+,F}| > \epsilon m$ and 0 otherwise, and let $Z = \sum_{F \in \mathcal{F}} Z_F$. We must prove $E[Z] \leq 1/2$. Fix $F \in \mathcal{F}$; it suffices to show $Pr[|M_{+,F}| > \epsilon m] \leq \frac{1}{2|\mathcal{F}|}$. For $i \in [m]$ define $X_{i,F} = M_{i,F} - E[M_{i,F}]$ and define $X_{+,F} = \sum_{i=1}^{n} X_{i,F} = M_{+,F} - E[M_{+,F}]$. We claim that $|M_{+,F}| > \epsilon m$ implies $|X_{+,F}| > (\epsilon - \delta)m$. To see this, note that $|E[M_{i,F}]| \leq \delta$ for all $i \in [m]$, since D' is everywhere δ-biased. Hence, if $|M_{+,F}| > \epsilon m$,

$$|X_{+,F}| = |M_{+,F} - E[M_{+,F}]| \geq |M_{+,F}| - |E[M_{+,F}]| \geq |M_{+,F}| - \sum_{i=1}^{m} |E[M_{i,F}]| > \epsilon m - \delta m.$$

Therefore to prove the claim, it suffices to show that $Pr[|X_{+,F}| > (\epsilon - \delta)m] \leq \frac{1}{2|\mathcal{F}|}$.

Now, it is straightforward to upper bound $Pr[|X_{+,F}| > \epsilon - \delta]$ using a Chernoff-type large deviation inequality. We follow an argument used in [AS91] (Corollary A.7); the details are needed in the next subsection. By definition of M, the random variables $M_{1,F}, \ldots, M_{m,F}$ are mutually independent, and hence so are $X_{1,F}, \ldots, X_{m,F}$. Defining $p_F = Pr[M_{i,F} = 1]$ which is independent of i and equal to $Pr[X_{i,F} = 1 - E[M_{i,F}]]$, we have $E[M_{i,F}] = 2p_F - 1$. Now, for any λ,

$$E[e^{\lambda X_{i,F}}] = p_F e^{\lambda(1 - E[M_{i,F}])} + (1 - p_F)e^{\lambda(-1 - E[M_{i,F}])}$$
$$= p_F e^{2\lambda(1-p_F)} + (1 - p_F)e^{-2\lambda p_F}. \tag{1}$$

The last expression can be bounded above by $e^{\lambda^2/2}$ by taking $\alpha = p_F$ and $\beta = 2\lambda$ in the following technical lemma from [AS91]:

Lemma 3.2 *For all $\alpha \in [0,1]$ and all β, $\alpha e^{\beta(1-\alpha)} + (1 - \alpha)e^{-\beta\alpha} \leq e^{\beta^2/8}$.*

Therefore,

$$E[e^{\lambda X_{+,F}}] = \prod_{i=1}^{m} E[e^{\lambda X_{i,F}}] \leq e^{m\lambda^2/2}. \tag{2}$$

Applying Markov's inequality, we have

$$Pr[X_{+,F} > a] = Pr[e^{\lambda X_{+,F}} > e^{\lambda a}] < \frac{E[e^{\lambda X_{+,F}}]}{e^{\lambda a}} \le e^{m\lambda^2/2 - \lambda a}.$$

The right hand side is minimized at $\lambda = a/m$ and hence $Pr[X_{+,F} > a] < e^{-a^2/2m}$. By symmetry, $Pr[|X_{+,F}| > a] < 2e^{-a^2/2m}$. Substituting $(\epsilon - \delta)m$ for a and using our choice of m, we have

$$Pr[|X_{+,F}| > (\epsilon - \delta)m] < \frac{1}{2|\mathcal{F}|}.$$

This completes the proof.   □

In particular, we have $m = O(\frac{\log n}{\epsilon^2})$ if $\mathcal{F}$ has size polynomial in n and if $\epsilon - \delta = \Theta(\epsilon)$.

3.2   The Explicit Construction

The previous lemma can be viewed as giving a randomized algorithm that with probability greater than $1/2$, constructs an $m \times n$ matrix that is ϵ-biased matrix on $\mathcal{F}$ with $m = \frac{2\ln(4|\mathcal{F}|)}{(\epsilon-\delta)^2}$. We now apply the method of conditional probabilities to derandomize this algorithm, as sketched in the introduction to this section. For definiteness we set $\delta = \epsilon/2$.

First we choose the distribution D'. By Theorem 2.1, there is an efficient construction of a $m' \times n$ matrix M' with $m' = poly(\frac{n}{\epsilon})$ that is $\epsilon/2$-biased.

The only remaining problem with applying the method of conditional probabilities here is that we do not have an efficient algorithm for computing (or estimating) the needed conditional probabilities for Z. So we construct a pessimistic estimator W for Z (essentially the same one used by Raghavan [Rag88] in a different context). For $F \in \mathcal{F}$, define the random variable:

$$W_F = e^{-(\epsilon-\delta)^2 m}\left(e^{(\epsilon-\delta)X_{+,F}} + e^{-(\epsilon-\delta)X_{+,F}}\right),$$

and define $W = \sum_{F \in \mathcal{F}} W_F$.

We first show that W satisfies conditions (A) and (B) of a pessimistic estimator. For condition (A) we must show $W(M) \ge Z(M)$ for any matrix M and for this it suffices to show that $W_F(M) \ge Z_F(M)$ for any $F \in \mathcal{F}$ and matrix M. $W_F(M)$ is always positive so by the definition of Z, it suffices to show that if $|X_{+,F}(M)| > (\epsilon - \delta)m$ then $W_F(M) \ge 1$, which is obvious from the definition of W_F.

For condition (B), we show that $E[W] < 1/2$ and for this it suffices that $E[W_F] < \frac{1}{2|\mathcal{F}|}$ for each $F \in \mathcal{F}$. Using the definition of W_F and (2) we have $E[W_F] \le 2e^{-(\epsilon-\delta)^2 m/2}$ and the desired inequality follows from the choice of m.

So W is a pessimistic estimator for Z, and hence can be used in place of Z when applying the method of conditional probabilities. Here we are in the situation described in Remark 2.1: we can only estimate the conditional probabilities to some desired precision. Since we have $E[W] < 1/2$ and we need to construct an M with $W(M) < 1$, it suffices to have precision $\mu = \frac{1}{2m}$. We claim that for any $i \in [m]$ and any choice $v^1, v^2, \ldots, v^i$ of the first i rows of M, we can estimate $E[W_F|v^1, v^2, \ldots, v^i] := E[W_F|M_{1,*} = v^1, \ldots, M_{i,*} = v^i]$ to within μ in time polynomial in n, $1/\epsilon$ and $1/\mu$. This expectation is a sum of two terms, and the computation for each is essentially the same, so we sketch how to estimate only the first term to within $\mu/2$, leaving the rest of the details to the reader.

By definition $X_{+,F} = \sum_{k=1}^{m} X_{k,F}$ and the $X_{k,F}$'s are mutually independent, so

$$E[e^{(\epsilon-\delta)X_{+,F}}|v^1, v^2, \ldots, v^i] = \prod_{k=1}^{m} E[e^{(\epsilon-\delta)X_{k,F}}|v^1, \ldots, v^i].$$

Assume without loss of generality that $\epsilon < 1/2$, this implies in particular that $e^{\epsilon-\delta} \leq 2$. Since $X_{k,F}$ takes values ± 1 we conclude that each term in the product is between $1/2$ and 2. Hence, to estimate the product within $\mu/2$ it suffices to estimate each term in the product with additive error $\frac{\mu}{m2^m}$. p_F is easy to compute given M': it is just the proportion of 1's in the vector $M'_{*,F}$. For each k, $X_{k,F} = M_{k,F} - E[M_{k,F}]$ depends only on the k-th row $M_{k,*}$ so for each $1 \leq k \leq i$,

$$E[e^{(\epsilon-\delta)X_{k,F}}|v_1, \ldots, v_i] = e^{(\epsilon-\delta)(M_{k,F}-(2p_F-1))}$$

which is easy to estimate using the Taylor expansion of e^x (here the argument x is at most $1/2$ so that we need expand it only $O(m)$ terms to get the required precision). For each $i+1 \leq k \leq m$,

$$E[e^{(\epsilon-\delta)X_k^J}|v_1, \ldots, v_i] = E[e^{(\epsilon-\delta)X_k^J}]$$

which by (1) is easy to estimate as well. □

To summarize, our algorithm is: first construct the matrix M'. Then build the matrix M row by row, selecting each successive row to be the row of M' that minimizes (to within μ) the conditional expectation of W given the rows already selected. We have shown:

Theorem 3.1 *There exists a deterministic algorithm that, on input integer n, family $\mathcal{F}$ of subsets of $[n]$ and $\epsilon \in (0,1)$, runs in time $poly(|\mathcal{F}|n/\epsilon)$ and outputs a $m \times n$ matrix M that is ϵ-biased on $\mathcal{F}$ with $m = O(\frac{\log|\mathcal{F}|}{\epsilon^2})$.*

In particular, in the case that $\mathcal{F}$ has size polynomial in n and $\epsilon = \frac{1}{poly(n)}$, the algorithm runs in $poly(n)$ time and the matrix M has $O(\frac{\log n}{\epsilon^2})$ rows.

Taking k to be a constant, and $\mathcal{F}$ to be the set of all subsets of size at most k, we get a polynomial time construction of a k-wise ϵ-biased matrix with $O(\log n/\epsilon^2)$ rows.

4 Applications

4.1 Error-Correcting Communication Protocols

In the standard model of two-party communication complexity [Yao79, KN95], party A has input x from some domain X, party B has input y from some domain Y and they must evaluate a given function $f(x,y)$. The goal is to do this with as little communication as possible. (Without loss of generality, we assume here that it suffices for either party to know the answer.) The cost of local computation is typically ignored. For any function, a naive protocol is to have A send x to B and have B compute $f(x,y)$ locally. Using an appropriate (binary) encoding of X, the communication needed for this $\log|X|$.

A basic example of a function f for which this problem is of interest is the *equality* function, $EQ(x,y) = 1$ if $x = y$ and 0 otherwise. It has been shown [Yao79] that for this problem, the naive protocol is optimal among deterministic protocols.

We assume $X = Y = \{0,1\}^n$. We consider a variant of EQUALITY in which the two parties may assume that if the two strings differ then their difference pattern is one of a predetermined set of patterns. Also we will not ignore the cost of local computation: we are interested in protocols that minimize communication while also keeping local computation small.

Notation and Definitions For a set $J \subseteq [n]$, define χ_J to be the characteristic vector of J, i.e., $\chi_J \in \{0,1\}^n$ and the i-th bit of χ_J is 1 if and only if $i \in J$.

In this section it will be convenient to switch our boolean set to $\{0,1\}$. The "+" operation will always be addition modulo 2. We translate the results of the previous section by mapping -1 to 1 and 1 to 0 and multiplication becomes addition mod 2. Thus if N is a $\{0,1\}$ matrix $N_{i,F}$ is the mod 2 sum of the entries in the ith row indexed by F. The condition that a $\{0,1\}$-matrix N is ϵ biased on the set F is that the fraction of 0's in the vector $N\chi_F$ (the mod 2 matrix-vector product) is between $(1-\epsilon)/2$ and $(1+\epsilon)/2$.

For two sets $J, J' \in [n]$, $J \triangle J'$ denotes the symmetric difference $(J - J') \cup (J' - J)$. The *difference pattern* of two vectors $x, y \in \{0,1\}^n$ is the vector $x + y$, and the *difference set*, $\Delta(x,y)$ is the set of positions on which they differ. Note that $\chi_{J \triangle J'} = \chi_J + \chi_{J'}$ and $\chi_{\Delta(x,y)} = x + y$.

The Communication Problem and the Approach We suppose that a family $\mathcal{F}$ of subsets of $[n]$, which we call the set of *expected differences*, is known to both parties beforehand. The $\mathcal{F}$-*error detection* is to compute $EQ(x,y)$ assuming that the inputs x and y satisfy $\Delta(x,y) \in \mathcal{F}$. In the $\mathcal{F}$-*error correction* problem A and B must also determine the error pattern $x + y$.

If $\mathcal{F}$ consists of $\Lambda([n])$, then the error detection problem is the same as EQUALITY. Therefore, for general $\mathcal{F}$, A and B can do nothing better than the naive protocol given above. However, for specific $\mathcal{F}$ this can be improved: we will present a deterministic protocol using $O(\log|\mathcal{F}|)$ bits of communication.

The amount of local computation required is $O(poly(|\mathcal{F}|))$. In particular, if the two parties are polynomial time machines the protocol allows them to handle the case that $\mathcal{F}$ has polynomial size with $O(\log n)$ bits of communication.

The bound of $O(\log|\mathcal{F}|)$ on communication is often tight. Consider the following two examples in the case of error detection.

Example 1: Suppose $\mathcal{F}$ includes $\Lambda\left(\binom{[n]}{2}\right)$. Restricting the input to vectors with a single 1, we get the equality problem for a domain of size n; as mentioned above there is a $\log n$ lower bound for this.

Example 2: For a subset $J \subseteq [n]$ of size $\Theta(\log n)$, we suppose that $\mathcal{F}$ includes $\Lambda(J)$. Restricting the input to strings that are 0 outside of J we have the equality problem for a domain of size $\text{poly}(n)$ which again has a $\Omega(\log n)$ lower bound.

In both of these examples, if $|\mathcal{F}| = n^{O(1)}$, our protocol uses $O(\log(|\mathcal{F}|)) = O(\log n)$ bits, which matches the lower bound within a constant factor.

The design of our protocol exploits the explicit construction of sample spaces with small bias on neighborhoods. Naor and Naor in [NN90] were the first to apply explicit constructions of everywhere ϵ-biased matrices to communication problems. In their paper, they presented a randomized protocol for the EQUALITY problem that achieves polynomially small error using $O(\log n)$ bits. Here is their protocol. Assume A and B each know a matrix N that is everywhere ϵ-biased. A picks a row N_i from N at random and sends the index i together with the bit $N_i \cdot x$ to B. B then computes $N_i \cdot y$ and decides that $x = y$ if and only if $N_i \cdot y = N_i \cdot x$. Note that B will never incorrectly say that $x \neq y$. By the ϵ bias of N, the fraction of 1's in $N(x + y)$ is at least $(1 - \epsilon)/2$ and so the probability that B incorrectly says $x = y$ is at most $\frac{1+\epsilon}{2}$. By independent repeated trials, they can then achieve arbitrarily high success probability.

For a deterministic protocol, on the other hand, A could send Nx to B and B would output 1 if and only if $Nx = Ny$. If $r(N)$ is the number of rows of N then the communication complexity is $\log r(N)$ for the randomized protocol and $r(N)$ for the deterministic protocol.

Let us now return to our problem. Consider first the $\mathcal{F}$-error-detection problem. The work in [NN90] provides an explicit construction of $(\Upsilon(\mathcal{F}))$-wise $\frac{1}{2}$-biased matrices with $\Theta(\Upsilon(\mathcal{F}) \log n)$ rows, which in our case is polylogarithmic in n. Such a matrix is by definition $\frac{1}{2}$-biased on $\mathcal{F}$. Therefore, the randomized protocol described above works for our problem and uses $O(\log\log n)$ bits of communication. The deterministic protocol uses $\Theta(\Upsilon(\mathcal{F}) \log n)$ bits. In particular, in the case that $\Upsilon(\mathcal{F})$ is $\omega(1)$, this deterministic protocol would use $\omega(\log n)$ bits.

By Theorem 3.1, we can construct a matrix $N_{\mathcal{F},1/2}$ that is ϵ-biased on $\mathcal{F}$ and has $O(\log|\mathcal{F}|)$ rows, and so we get a deterministic protocol for the error detection problem for $\mathcal{F}$ with $O(\log n)$ communication. Note that to implement this protocol, we have assumed that the two parties have a common matrix $N_{\mathcal{F},1/2}$. Our algorithm provides a way for them to construct this matrix efficiently.

In the next subsection we extend this idea to get a deterministic error-correcting protocol.

The Protocol For a family $\mathcal{F}$ of sets, we define $\mathcal{F}^* = \{X \triangle Y \mid X, Y \in \mathcal{F}\}$. Clearly $|\mathcal{F}^*| = O(|\mathcal{F}|^2)$.

For $\mathcal{F}$-error correction A and B want to determine the difference set $\Delta(x, y)$, under the assumption that $\Delta(x, y) \in \mathcal{F}$. We assume that A and B each know (the same) matrix N that is $1/2$-biased on $\mathcal{F}^*$, which can be constructed using the algorithm from the previous section. If J, J' are distinct sets of $\mathcal{F}$ then at least $1/4$ of the entries of $N\chi_{J \triangle J'}$ are 1, and in particular $N\chi_J \neq N\chi_{J'}$. Hence the function that maps each $J \in \mathcal{F}$ to the vector $N\chi_J$ is one-to-one; let V denote its image and $g : V \longrightarrow \mathcal{F}$ its inverse. We assume that A and B keep a table of the function g. The time to construct N and the table is polynomial in $|\mathcal{F}^*|$ which is polynomial in $|\mathcal{F}|$. The protocol is:

1. A sends Nx to B.
2. B computes $z = N(x + y) = Nx + Ny$ and determines $\Delta(x, y)$ to be $g(z)$.

The protocol clearly uses $O(\log n)$ bits of communication and its correctness is immediate from the definition of the function g.

4.2 Constructing $(n, \mathcal{F})$-universal Sets

If $T \subseteq \{0, 1\}^n$ and $F \subseteq [n]$, we say that T *shatters* F if the restriction of T to the coordinates indexed by F contains $\{0, 1\}^{|F|}$. If $\mathcal{F}$ is a collection of subsets of $[n]$ we say that T is $(n, \mathcal{F})$-*universal* if T shatters each $F \in \mathcal{F}$. Such sets were considered in the context of testing logic circuits (see [SB88]). In this subsection, using our construction of sample spaces with small bias on neighborhoods, we prove:

Theorem 4.1 *Let $\mathcal{F}$ be a family of subsets of $[n]$. Then there is an explicit construction of an $(n, \mathcal{F})$-universal set of size $O(2^{\Upsilon(\mathcal{F})} \log |\Lambda(\mathcal{F})|)$. The construction uses time polynomial in n and $|\Lambda(\mathcal{F})|$.*

To prove the theorem we need the following fact from [NN90] (whose proof is based on a lemma from [Vaz86]):

Proposition 4.1 *If N is a $0, 1$ matrix with k columns that is 2^{-k-1}-biased, then every vector in $\{0, 1\}^n$ appears as a row of N.*

As an immediate consequence of this and the definition of $(n, \mathcal{F})$-universalwe have:

Corollary 4.1 *If N is a $0, 1$ matrix with n columns that is $2^{-\Upsilon(\mathcal{F})-1}$-biased with respect to $\Lambda(\mathcal{F})$, then the set T consisting of distinct rows of N is $(n, \mathcal{F})$-universal.*

The corollary and Theorem 3.1 give us Theorem 4.1.

Schulman [Sch92] gave an explicit construction of $(n, \mathcal{F})$-universal sets of size $O(d2^{\Upsilon(\mathcal{F})})$, where d denotes the maximum number of neighborhoods containing any particular element. In general, neither construction is better than the other.

Theorem 4.1 in particular gives a polynomial time construction of an $(n, \binom{[n]}{k})$-universal set of size $O(2^k k \log n)$ for fixed k, which is optimal following the lower bound argument in [SB88]. Alon [Alo86] showed an explicit construction of $(n, \binom{[n]}{k})$-universal sets of size $\log n \cdot 2^{O(k^2)}$ for the case k is fixed. In [NN90] and [ABNNR92], nearly optimal constructions of size $\log n \cdot 2^{O(k)}$ were given. The previously best known construction in this case is due to [NSS95] and has size $2^k k^{O(\log k)} \log n$. Our construction has the least size and at the same time matches the bound given by the probabilistic argument.

References

[Alo86]   N. Alon, *Explicit constructions of exponential sized families of k-independent sets*, Discrete Math, 58, pp 191-193, 1986.

[Alo91]   N. Alon, *A parallel algorithm version of the Local Lemma*, Random Structures and Algorithms 2(4), pp 367-378, 1991.

[ABI86]   N. Alon, L. Babai, A. Itai, *A fast and simple randomized parallel algorithm for the Maximal Independent Set Problem*, J. Algorithms 7, pp 567-583, 1986.

[ABNNR92] N. Alon, J. Bruck, J. Naor, M. Naor, R. Roth, *Construction of asymptotically good low-rate error-correcting codes through pseudo-random graphs*, IEEE Trans. on Info. Theory, Vol 38(2), pp 509-515, 1992.

[AGHP91]   N. Alon, O, Goldreich, J. Hastad, R. Peralta, *Simple constructions of almost k-wise independent random variables*, Random Structures and Algorithms 3(3), pp 289-303, 1992.

[AS91]   N. Alon, J. Spencer, *The Probabilistic Method*, Wiley, 1991.

[BR89]   B. Berger, J. Rompel, *Simulating $(\log^c n)$-wise independence in NC*, J. Assoc. Comput. Mach. 38(4), pp 1026-1046, 1991.

[CGHFRS85] B. Chor, O. Goldreich, J. Hastad, J. Friedman, S. Rudich, R. Smolensky, *The bit extraction problem or t-resilient functions*, 26th IEEE FOCS, pp 396-407, 1985.

[CRS94]   S. Chari, P. Rohatgi, A. Srinivasan, *Improved algorithms via approximations of probability distributions*, 26th ACM STOC, pp 584-592, 1994.

[EGLNV92] G. Even, O. Goldreich, M. Luby, N. Nisan, B. Velickovic, *Approximations of general independent distributions*, 24th ACM STOC, pp 10-16, 1992.

[ES73]   P. Erdös, J. Selfridge, *On a combinatorial game*, Journal of Combinatorial Theory, Series A 14, pp 298-301, 1973.

[Fun94]   A. Fundia, *Derandomizing Chebyshev's inequality to find independent sets in uncrowded hypergraghs*, Random Structures and Algorithms, to appear.

[KK94]   D. Karger and D. Koller, *(De)randomized construction of small sample spaces in NC*, 35th IEEE FOCS, pp 252-263, 1994.

[KM94]   H. Karloff and Y. Mansour, *On construction of k-wise independent random variables*, Proceedings of the 26th Annual ACM Symposium on Theory of Computing, 1994.

[KW84]   R. Karp and A. Wigderson, *A fast parallel algorithm for the maximal independent set problem*, Proceedings of the 16th Annual ACM Symposium on Theory of Computing, 1984.

[KM93]   D. Koller and N. Megiddo, *Finding small sample spaces satisfying given constraints*, Proceedings of the 25th Annual ACM Symposium on Theory of Computing, pp 268-277, 1993.

[KN95]   Kuselevitz and N. Nisan, *Communication Complexity*.

[LLSZ95]   N. Linial, M. Luby, M. Saks, D. Zuckerman, *Efficient construction of a small hitting set for combinatorial rectangles in high dimension*, 25th ACM STOC, pp 258-267, 1993.

[Lub85]   M. Luby, *A simple parallel algorithm for the maximal independent set problem*, SIAM J. Comput. 15(4), pp 1036-1053, 1986.

[MNN89]   R. Motwani, J. Naor, M. Naor, *The probabilistic method yields deterministic parallel algorithms*, 30th IEEE FOCS, pp 8-13, 1989.

[NN90]   J. Naor, M. Naor, *Small-bias probability spaces: efficient constructions and applications*, SIAM J. Comput. 22(4), pp 838-856, 1993.

[NSS95]   M. Naor, L. Schulman, A. Srinivasan, *Splitters and near-optimal derandomization*, 36th IEEE FOCS, pp 182-191, 1995.

[Rag88]   P. Raghavan, *Probabilistic construction of deterministic algorithms: approximating packing integer programs*, JCSS 37, pp 130-143, 1988.

[Sch92]   L. Schulman, *Sample spaces uniform on neighborhoods*, 24th ACM STOC, pp 17-25, 1992.

[SB88]   G. Seroussi, N. Bshouty, *Vector sets for exhaustive testing of logic circuits*, IEEE Trans. on Info. Theory, Vol 34(3), pp 513-522, 1988.

[Spe87]   J. Spencer, *Ten Lectures on the Probabilistic Method*, SIAM(Philadelphia), 1987.

[Vaz86]   U. Vazirani, *Randomness, Adversaries and Computation*, Ph.D. Thesis, University of California, Berkeley, 1986.

[Yao79]   A. C-C Yao, *Some complexity questions related to distributive computing*, 11th ACM STOC, pp 209-213, 1979.

Approximation on the Web: A Compendium of NP Optimization Problems

P. Crescenzi[1][*] and V. Kann[2]

[1] Dipartimento di Scienze dell'Informazione
Università degli Studi di Roma "La Sapienza"
Via Salaria 113, 00198 Rome, Italy
E-mail: piluc@dsi.uniroma1.it
[2] Department of Numerical Analysis and Computing Science
Royal Institute of Technology
S-100 44 Stockholm, Sweden
E-mail: viggo@nada.kth.se

Abstract. A compendium of NP optimization problems, containing the best approximation results known for each problem, is now available on the web at http://www.nada.kth.se/~viggo/problemlist/. In this paper we describe such a compendium, and specify how the compendium is consultable (and modifiable) on the world wide web.

1 Introduction

The last few years have seen an extremely prolific research activity in the field of design and analysis of approximation algorithms. Actually, the notion of approximation algorithm has been considered since the very beginning of the theory of NP-completeness as an alternative way of coping with the difficulty of solving NP-hard combinatorial optimization problems. In 1974 Johnson [13] proposed several approximation algorithms for significant combinatorial problems such as MAXIMUM SATISFIABILITY, MINIMUM SET COVER, and MAXIMUM CLIQUE.

It is worth observing that Johnson's algorithms have been the best known ones for the above problems until the beginning of the 90s. This fact can, in our opinion, be justified by the following three arguments. First, some of them are *indeed* the best possible algorithms (as it has recently been proved in the case of MINIMUM SET COVER [6] and in the case of MAXIMUM k-SATISFIABILITY [10]). Second, designing better approximation algorithms turned out to be a difficult task that sometimes required more sophisticated algorithmic techniques (as in the case of MAXIMUM SATISFIABILITY [9]). Third, researchers have probably not devoted enough attention to this field until a breakthrough paper by Papadimitriou and Yannakakis appeared in 1989 [18].

---

[*] Starting from November 1, 1997, the author's new affiliation will be: Dipartimento di Sistemi ed Informatica, Università di Firenze, Via Cesare Lombroso 6/17, 50134 Firenze, Italy (e-mail: piluc@dsi2.dsi.unifi.it).

That paper did not present new approximation algorithms but it put in a common framework many interesting approximation results that were known so far and proposed in a different way the same open problems that were already explicitly stated in Johnson's paper. As a result, the field of approximation algorithms received attention by the community working in the field of interactive proof systems and of probabilistically checkable proofs which at a first glance could seem to be unrelated. The sequence of results that finally led to the so-called *PCP-theorem* has thus made it possible to prove some important non-approximability results, such as the fact that MAXIMUM SATISFIABILITY does not admit a polynomial-time approximation scheme.

Since then, the number of both positive and negative results regarding the approximability properties of NP-hard combinatorial problems has continuously increased with an "exponential" growth rate. And it does not seem to decline. Currently, there is basically no conference or workshop in the field of algorithms and complexity that does not include in its proceedings at least two or three papers regarding the design of approximation algorithms. As a consequence, the number of problems for which approximation algorithms and/or non-approximability results have been obtained has drastically increased in the last six years. Moreover, the degree of approximability of several important problems (mostly contained in the appendix of the book by Garey and Johnson [8]) has changed very rapidly. One single example: the MAXIMUM 2-SATISFIABILITY started from the 2-approximation algorithm of Johnson and in a few years reached the incredible 1.0741-approximation algorithm of Feige and Goemans [7] (which is almost the best possible performance ratio obtainable for this problem [10]).

Following this rapid development is thus becoming a hard task. This is the main motivation for collecting the existing approximation results into a compendium available to all researchers working (or starting to work) in this field. This paper describes such a compendium, and specifies how the compendium is consultable (and modifiable) on the world wide web.

2   A compendium of NP optimization problems

In 1993 the two authors of this paper started to collect approximation results. We soon noticed that the amount of approximability results was enormous, and therefore we had to limit ourselves to consider polynomial-time approximation results for NP optimization problems that are not solvable optimally in polynomial time. This means that we considered neither approximation algorithms for #P-hard or PSPACE-hard problems, nor fast approximation algorithms for polynomially solvable problems such as MAXIMUM MULTICOMMODITY FLOW. We of course collected the best upper and lower bounds of approximation, but we did not try to find the fastest approximation algorithm giving the best approximation upper bound – any polynomial time algorithm sufficed.

With these restrictions, in 1994 we made the first version of the compendium available as a Postscript file via anonymous ftp, and announced it in the news group comp.theory. The response from the research community was fortunately enough extensive, and we received both corrections, updated results and results for new problems. Since then we have made several revisions of the compendium, and the current version has number six. In Section 5 we explain how you can help us to improve the compendium.

We noted early that a Postscript file is not the ideal medium for the compendium. A researcher is often only interested in looking up the results for a single problem. In the end of 1994 we therefore constructed a web version of the compendium, containing exactly the same text as the Postscript file, but with hypertext links between the different sections and problems, making it a lot easier to use. In Section 4 some statistics of the use are presented.

2.1 Other optimization problem lists

The book by Garey and Johnson [8] does not just contain a comprehensive list of NP-complete problems, but also a chapter with upper and lower bounds for the approximability of NP optimization problems.

A first attempt to construct a problem list of approximability results for NP optimization problems was made by Kann [14]. This list was in fact the foundation on which the compendium described in this paper was built.

At the end of the recent book edited by Hochbaum [11] there is an alphabetical list of optimization problems (NP optimization problems as well as both easier and harder problems). However, there are no approximability results in the list, only pointers to the section in the book where the problem is mentioned.

3 How is the compendium structured?

We have chosen to structure the problems in the same way as Garey and Johnson did [8], that is systematically in twelve categories according to subject matter. The first five categories are divided into subcategories. The following table shows how many problems there are in each category in the April 1997 version of the compendium, compared to how many there are in [8].

Code	Name of category	Number of problems in	
		G&J [8]	compendium
GT	Graph theory	65	51
ND	Network design	51	59
SP	Sets and partitions	21	11
SR	Storage and retrieval	36	10
SS	Sequencing and scheduling	22	20
MP	Mathematical programming	13	17
AN	Algebra and number theory	18	1
GP	Games and puzzles	15	2
LO	Logic	19	13
AL	Automata and language theory	21	5
PO	Program optimization	10	1
MS	Miscellaneous	19	14

For many categories the numbers are about the same: moreover, under the same basic problem, several variations are also included. This might be surprising considering that just some of the NP-complete problems in [8] have corresponding optimization problems. The explanation is of course that several new NP-complete problems have been studied since 1979: an updated list of NP-complete problems would be much larger than the original one.

It is interesting that, in some categories, there are much fewer problems in the compendium than in [8]. The reason could either be that there are only few optimization problems in these categories or that no approximability results have been shown for the optimization problems in these classes. The latter case suggests that more research is needed for these categories.

The current version of the compendium contains more than 200 problems. A typical entry consists of eight parts: the first four parts are mandatory while the last four parts are optional.

1. The problem name that also specifies the goal of the problem.
2. The definition of the instances of the problem.
3. The definition of the feasible solutions of the problem.
4. The definition of the measure of a feasible solution.
5. A 'good news' part that contains the best approximation positive result (upper bound) for the problem.
6. A 'bad news' part that contains the worst approximation negative result (lower bound) for the problem.
7. A section of additional comments. In this section approximability results for variations of the problem are mentioned.
8. A reference to the 'closest' problem appearing in the list published in [8].

The following is an example of what an entry looks like.

GT7. MINIMUM EDGE COLORING

INSTANCE: Graph $G = \langle V, E \rangle$.
SOLUTION: A coloring of E, that is, a partition of E into disjoint sets $E_1, E_2, \ldots, E_k$ such that, for $1 \le i \le k$, no two edges in E_i share a common endpoint in G.
MEASURE: Cardinality of the coloring, i.e., the number of disjoint sets E_i.

Good News: Approximable within 4/3, and even approximable with an absolute error guarantee of 1 [20].
Bad News: Not approximable within $4/3 - \varepsilon$ for any $\varepsilon > 0$ [12].
Comment: Also called *Minimum Chromatic Index*. APX-intermediate unless the polynomial-hierarchy collapses [5]. On multigraphs the problem is approximable within $1.1 + (0.8/opt)$ [15]. The maximization variation in which the input is extended with a positive integer k, and the problem is to find the maximum number of consistent vertices over all edge-colorings with k colors, is approximable within $e/(e-1)$ [4], but does not admit a PTAS [19].
Garey and Johnson: OPEN5

3.1 Preliminaries

In order to consult the compendium, some preliminary definitions and results regarding approximation algorithms must be known. These definitions can be found in several textbooks on complexity theory such as [8, 17, 3, 16] or on approximation theory such as [11, 2]. The compendium itself is preceded by an introduction (also available on the web) that contains all necessary definitions and preliminary results.

4 How is the compendium used today?

The web version of the compendium on URL

```
http://www.nada.kth.se/~viggo/problemlist/
```

was in April 1997 used about 40 times each day, and in each session about 7 web pages were accessed. During the month more than 400 users from 50 different countries around the world accessed the web compendium.

Figure 1 shows how much the compendium has been used from different countries.

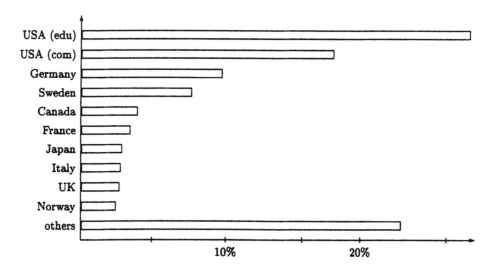

Fig. 1. Use of the compendium by country.

5 The future of the compendium

The above numbers show that the compendium seems to be considered very useful by researchers and encourage us to continue in mantaining it. Updating and completing the compendium, however, is an endless work, so there will never be a final version of the compendium. The compendium will be part of a new book on approximation [2], but the web version of the compendium will continue to evolve after the book is printed.

It is impossible for the authors to keep up with all new approximation results presented at conferences and published in journals without help from others. In the future we will trust that any researcher publishing a new result that would fit in the compendium will report it to us. In order to facilitate this communication we have created three web forms for

1. entering results for a problem that already is in the compendium,
2. entering results for a new problem,
3. reporting errors.

The forms are available from the home page of the compendium and designed to be easy to use. Figure 2 is an example of the first form.

6 Conclusion

In this paper we have briefly described a compendium of NP optimization problems which is now available on the web. We believe that such a compendium will turn out to be very useful whenever someone has to deal with the approximate solution of an NP-hard optimization problem. Indeed, as stated in [1], "the first step in proving an inapproximability result for a given problem is to check whether it is already known to be inapproximable." The compendium is then the right starting point to performe this test (actually, this is true also for positive results). Moreover, even though the problem at hand is not contained in the compendium, it is likely that it contains other problems which can be reduced to the problem at hand.

We have also described how the compendium can now be updated directly on the web by means of three forms which have been designed to be easy to use. We hope that in this way researchers will be stimulated to help us in maintaining the compendium as up-to-date as possible: the sooner they report a result the sooner it will be publicly known!

References

1. Arora, S., and Lund, C. (1997), *Hardness of approximations*, chapter 10 of [11]. PWS, 1997.
2. Ausiello, G., Crescenzi, P., Gambosi, G., Kann, V., Marchetti Spaccamela, A., and Protasi, M. (1997), *Approximate solution of NP-hard optimization problems*. To appear (a preliminary version of some chapters is available on request to one of the authors).

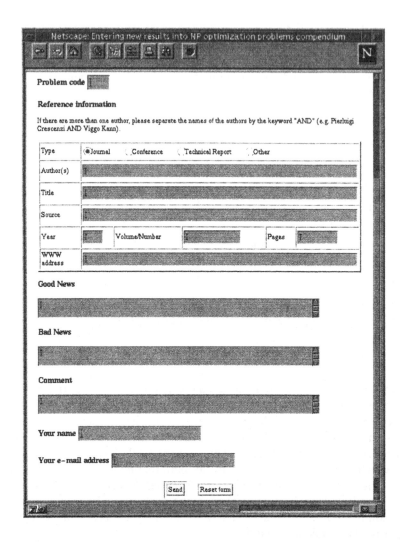

Fig. 2. The form for entering results for a problem already in the compendium

3. Bovet, D., and Crescenzi, P. (1993), *Introduction to the Theory of Complexity.* Prentice Hall.
4. Bertsimas, D., Teo, C-P., and Vohra, R. (1996), "On dependent randomized rounding algorithms", *Proc. 5th Int. Conf. on Integer Prog. and Combinatorial Optimization*, Lecture Notes in Comput. Sci. 1084, Springer-Verlag, 330–344.
5. Crescenzi, P., Kann, V., Silvestri, R., and Trevisan, L. (1995), "Structure in approximation classes", *Proc. 1st Ann. Int. Conf. on Computing and Combinatorics*, Lecture Notes in Comput. Sci. 959, Springer-Verlag, 539–548.
6. Feige, U. (1996), "A threshold of $\ln n$ for approximating set cover", *Proc. 28th Ann. ACM Symp. on Theory of Comp.*, ACM, 314–318.

7. Feige, U., and Goemans, M. X. (1995), "Approximating the value of two prover proof systems, with applications to MAX 2SAT and MAX DICUT", *Proc. 3rd Israel Symp. on Theory of Computing and Systems*, IEEE Computer Society, 182–189.

8. Garey, M. R., and Johnson, D. S. (1979), *Computers and Intractability: a guide to the theory of NP-completeness*, W. H. Freeman and Company, San Francisco.

9. Goemans, M. X., and Williamson, D. P. (1995b), "Improved approximation algorithms for maximum cut and satisfiability problems using semidefinite programming", *J. ACM* **42**, 1115–1145.

10. Håstad, J. (1997), "Some optimal inapproximability results", *Proc. 29th Ann. ACM Symp. on Theory of Comp.*, ACM, 1–10.

11. Hochbaum, D. S. ed. (1997), *Approximation algorithms for NP-hard problems*, PWS Publishing Company, Boston.

12. Holyer, I. (1981), "The NP-completeness of edge-coloring", *SIAM J. Comp.* **10**, 718–720.

13. Johnson, D. S. (1974), "Approximation algorithms for combinatorial problems", *J. Comput. System Sci.* **9**, 256–278.

14. Kann, V. (1992b), *On the Approximability of NP-complete Optimization Problems*, PhD thesis, Department of Numerical Analysis and Computing Science, Royal Institute of Technology, Stockholm.

15. Nishizeki, T., and Kashiwagi, K. (1990), "On the 1.1 edge-coloring of multigraphs", *SIAM J. Disc. Math.* **3**, 391–410.

16. Papadimitriou, C.H. (1994), *Computational Complexity*. Addison-Wesley.

17. Papadimitriou, C.H., and Steiglitz, K. (1982), *Combinatorial optimization: algorithms and complexity*, Prentice Hall.

18. Papadimitriou, C. H., and Yannakakis, M. (1991), "Optimization, approximation, and complexity classes", *J. Comput. System Sci.* **43**, 425–440.

19. Petrank, E. (1994), "The hardness of approximation: gap location", *Computational Complexity* **4**, 133–157.

20. Vizing, V. G. (1964), "On an estimate of the chromatic class of a p-graph", *Diskret. Analiz.* **3**, 23–30.

Random–Based Scheduling

New Approximations and LP Lower Bounds

(Extended Abstract)

Andreas S. Schulz and Martin Skutella

Technische Universität Berlin,
Fachbereich Mathematik, MA 6–1,
Straße des 17. Juni 136, 10623 Berlin, Germany,
E-mail {schulz,skutella}@math.tu-berlin.de

Abstract. Three characteristics encountered frequently in real-world machine scheduling are jobs released over time, precedence constraints between jobs, and average performance optimization. The general constrained one-machine scheduling problem to minimize the average weighted completion time not only captures these features, but also is an important building block for more complex problems involving multiple machines.

In this context, the conversion of preemptive to nonpreemptive schedules has been established as a strong and useful tool for the design of approximation algorithms. The preemptive problem is already NP-hard, but one can generate good preemptive schedules from LP relaxations in time-indexed variables. However, a straightforward combination of these two components does not directly lead to improved approximations. By showing schedules in slow motion, we introduce a new point of view on the generation of preemptive schedules from LP-solutions which also enables us to give a better analysis.

Specifically, this leads to a randomized approximation algorithm for the general constrained one-machine scheduling problem with expected performance guarantee e. This improves upon the best previously known worst-case bound of 3. In the process, we also give randomized algorithms for related problems involving precedences that asymptotically match the best previously known performance guarantees.

In addition, by exploiting a different technique, we give a simple $3/2$-approximation algorithm for unrelated parallel machine scheduling to minimize the average weighted completion time. It relies on random machine assignments where these random assignments are again guided by an optimum solution to an LP relaxation. For the special case of identical parallel machines, this algorithm is as simple as the one of Kawaguchi and Kyan [KK86], but allows for a remarkably simpler analysis. Interestingly, its derandomized version actually is the algorithm of Kawaguchi and Kyan.

1 Introduction

The main results of this paper are twofold. First, we give an approximation algorithm for the general constrained single machine scheduling problem to minimize the average weighted completion time. It has performance guarantee e whereas the best previously known worst-case bound is 3 [Sch96]. Second, we present another approximation algorithm for the model with unrelated parallel machines (but with independent jobs without non-trivial release dates) that has performance guarantee $3/2$. Previously, a 2–approximation algorithm was known [SS97]. Our first contribution is based on and motivated by earlier work of Hall, Shmoys, and Wein [HSW96], Goemans [Goe96,Goe97], and Chekuri, Motwani, Natarajan, and Stein [CMNS97]; the second one builds on earlier research by the authors [SS97]. All this work was in turn initiated by a paper of Phillips, Stein, and Wein [PSW95].

The clamp spanning our two main results is the use of randomness which in both cases is guided by optimum solutions to related LP relaxations of these problems. Hence, our algorithms actually are randomized approximation algorithms. A randomized ρ–approximation algorithm is an algorithm that produces a feasible solution whose expected value is within a factor of ρ of the optimum; ρ is also called the expected performance guarantee of the algorithm. However, all algorithms given in this paper can be derandomized with no difference in performance guarantee, but at the cost of increased running times. For reasons of brevity, most often we omit the technical details of derandomization.

In the first part, we consider the following model. We are given a set J of n jobs (or tasks) and a disjunctive machine (or processor). Each job j has a positive integral processing time p_j and an integral release date $r_j \geqslant 0$ before which it is not available. In *preemptive* schedules, a job may repeatedly be interrupted and continued at a later point in time. We generally assume that these preemptions only occur at integer points in time. In *nonpreemptive* schedules, a job must be processed in an uninterrupted fashion. We denote the completion time of job j in a schedule by C_j. In addition, $C_j(\alpha)$ for $0 < \alpha \leqslant 1$ denotes the earliest point in time when an α–fraction of j has been completed, in particular, $C_j(1) = C_j$; α–points were first used in the context of approximation by [HSW96]. The starting time of j is denoted by $C_j(0^+)$. We also consider precedence constraints between jobs. If $j \prec k$ for $j, k \in J$, it is required that j is completed before k can start, i. e., $C_j \leqslant C_k(0^+)$. We seek to minimize the average weighted completion time in this setting: a weight $w_j \geqslant 0$ is associated with each job j and the goal is to minimize $\frac{1}{n} \sum_{j \in J} w_j C_j$, or, equivalently, $\sum_{j \in J} w_j C_j$. In scheduling, it is quite convenient to refer to the respective problems using the standard classification scheme of Graham et al. [GLLRK79]. Both problems $1 \mid r_j, prec \mid \sum w_j C_j$ and $1 \mid r_j, pmtn, prec \mid \sum w_j C_j$ just described are strongly NP-hard. In the second part of this paper, we are given m unrelated parallel machines instead of a single machine. Each job j has a positive integral processing requirement p_{ij} which depends on the machine i job j will be processed on. Each job j must be processed for the respective amount of time on one of the m machines, and may be assigned to any of them. Every machine can process at most one job at a time. In standard notation, this NP-hard problem reads $R \mid \mid \sum w_j C_j$.

Chekuri, Motwani, Natarajan, and Stein [CMNS97] give a strong result about converting preemptive schedules to nonpreemptive schedules on a single machine. Consider any preemptive schedule with completion times C_j, $j = 1, \dots, n$. Chekuri et al. show that if α is selected at random in $(0, 1]$ with density function $e^\alpha/(e-1)$, then the expected completion time of job j in the schedule produced by sequencing the jobs in nondecreasing order of α–points is at most $\frac{e}{e-1} C_j$. This is a deep and in a sense best possible result. However, in order to get in this manner polynomial-time approximation algorithms for nonpreemptive single machine scheduling problems, one relies on good (exact or approximate) solutions to the respective preemptive version of the problem on hand. The only case where this immediately led to an improved performance guarantee was single machine scheduling with release dates (no precedence constraints) to minimize the average completion time (unit weights). We therefore suggest to convert so-called fractional schedules obtained from certain LP relaxations to obtain provably good nonpreemptive schedules for problems with precedence constraints and arbitrary (non-negative) weights. The LP relaxation we exploit is weaker than the one of Hall, Shmoys, and Wein [HSW96] which they used to derive the first constant-factor approximation algorithm for $1 | r_j, prec | \sum w_j C_j$. In fact, in contrast to their LP, our LP already is a relaxation of the corresponding preemptive problem $1 | r_j, pmtn, prec | \sum w_j C_j$. Hence, we are also able to give an approximation algorithm for the latter problem. In addition, we can interpret an LP solution as a preemptive schedule, but with preemptions at fractional points in time; schedules of this kind are called *fractional*. Our way of deriving and analyzing good fractional schedules from LP solutions generalizes the techniques of Hall, Shmoys, and Wein [HSW96] to this weaker LP relaxation. A somewhat intricate combination of these techniques with the conversion procedure of Chekuri et al. leads then to approximation algorithms for nonpreemptive single machine scheduling which improve or asymptotically match the best previously known algorithms. Specifically, for single machine scheduling with precedence constraints and release dates so as to minimize the average weighted completion time, we obtain in this way an e–approximation algorithm. The approximation algorithm of Hall, Shmoys, and Wein [HSW96] has performance guarantee 5.83. Inspired by the work of Hall et al., Schulz [Sch96] gave a 3–approximation algorithm based on a different LP.

Our second technique exploits optimum solutions to LP relaxations of parallel machine problems in a different way. The key idea, which has been introduced in [SS97], is to interpret the value of certain LP variables as the probability with which jobs are assigned to machines. For the quite general model of unrelated parallel machines to minimize the average weighted completion time, we show by giving an improved LP relaxation that this leads to a $3/2$–approximation algorithm. The first constant-factor approximation algorithm for this model was also obtained by Hall, Shmoys, and Wein [HSW96] and has performance guarantee $16/3$. This was subsequently improved to 2 [SS97]. One appealing aspect of this technique is that in the special case of identical parallel machines the derandomized version of our algorithm coincides with the algorithm of Kawaguchi and Kyan [KK86]; we therefore get a simple proof that list-scheduling in order of nonincreasing ratios of weight to processing time is a $3/2$–approximation.

The paper is organized as follows. In Sect. 2, we present the first technique in a quite general framework. It is best understood by showing schedules in slow motion. Then, in Sect. 3, we apply this to the general constrained one-machine scheduling problem. Finally, in Sect. 4, we randomly assign jobs to parallel machines.

2 Showing Schedules in Slow Motion

In addition to the setting described above, we consider instances with *soft* precedence constraints which will be denoted by $\prec'$. For $j, k \in J$, $j \prec' k$ requires that $C_j(\alpha) \leqslant C_k(\alpha)$

for $0 < \alpha \leqslant 1$. That is, at each point in time the completed fraction of job k must not be larger than the completed fraction of its predecessor j.

Of course, if we consider a single machine, it only makes sense to talk about soft precedence constraints if we allow preemption. However, even for the preemptive case the step from strong to soft precedence constraints is not a true relaxation for a single machine. This can be seen by considering the following algorithm:

Algorithm SOFT–TO–STRONG
Input: fractional schedule S that obeys release dates and soft precedence constraints.
Output: preemptive schedule that obeys release dates and strong precedence constraints.

sort: Sort the jobs in order of nondecreasing completion times in the given schedule S.

schedule: Construct a new schedule by scheduling at any point in time the first available job in this list.

Notice that Algorithm SOFT–TO–STRONG usually produces a preemptive schedule. Whenever a job is released, the job being processed (if any) will be preempted if the released job has smaller completion time in S. An example for Algorithm SOFT–TO–STRONG is given in the last two rows of Fig. 2.

Lemma 1. *Given a fractional schedule to an instance of* $1 \mid r_j, pmtn, prec' \mid \sum w_j C_j$, *Algorithm* SOFT–TO–STRONG *constructs in time* $O(n \log n)$ *a preemptive schedule obeying the corresponding strong precedence constraints and release dates with no increase in completion times.*

Proof. We suspect the lemma to belong to the folklore of this field; since we could not explicitly find it in the literature, however, we provide a proof for the sake of completeness. The reasoning is quite similar to the one given in [HSSW96, Proof of Lemma 2.8] for a rather different result. W. l. o. g., we assume $r_j \leqslant r_k$ whenever $j \prec k$ for $j, k \in J$. We denote the completion time of a job j in the given schedule S by C'_j and in the constructed schedule by C_j.

By construction, no job is processed before its release date in the new schedule. Moreover, a job is only preempted if another job is released at that time. Therefore, since all the release dates are integral, Algorithm SOFT–TO–STRONG creates preemptions only at integral points in time.

Furthermore, for $j \prec k$ the feasibility of the given schedule yields $C'_j < C'_k$. Thus, because of $r_j \leqslant r_k$ and our ordering of jobs, k is not processed before j is completed and the strong precedence constraints are obeyed.

It remains to show that $C_j \leqslant C'_j$ for each job $j \in J$. Let $t \geqslant 0$ be the earliest point in time such that there is no idle time in $(t, C_j]$ and only jobs k with $C'_k \leqslant C'_j$ are processed on the machine in the period from t to C_j in the new schedule. We denote the set of these jobs by K. By construction $r_k \geqslant t$ for all $k \in K$ and thus $C'_j \geqslant t + \sum_{k \in K} p_k$. On the other hand, the definition of K implies $C_j = t + \sum_{k \in K} p_k$ and therefore $C_j \leqslant C'_j$.

Finally, the running time of Algorithm SOFT–TO–STRONG is dominated by the sorting step and is therefore $O(n \log n)$. $\qquad\square$

If all the release dates are 0 the schedule constructed in Algorithm SOFT–TO–STRONG is nonpreemptive. Thus, we get the following corollary.

Corollary 2. *Given a fractional schedule to an instance of* $1 | pmtn, prec' | \sum w_j C_j$, *Algorithm* SOFT–TO–STRONG *computes in time* $O(n \log n)$ *a nonpreemptive schedule obeying the precedence constraints with no increase in completion times.*

It follows that the NP-hard problem $1 | prec | \sum w_j C_j$ [Law78] can be reduced to the problem $1 | pmtn, prec' | \sum w_j C_j$ which is therefore NP-hard, too. In order to develop approximation algorithms for all these problems we now introduce an LP relaxation of $1 | r_j, pmtn, prec' | \sum w_j C_j$ whose optimum value serves as a surrogate for the true optimum in our estimations. The LP relaxation is an immediate extension of a time-indexed LP proposed by Dyer and Wolsey [DW90] for the problem without precedence constraints. Here, time is discretized into the periods $(t, t+1]$, $t = 0, 1, \ldots, T$ where T is the planning horizon, say $T := \max_{j \in J} r_j + \sum_{j \in J} p_j - 1$. We have a variable y_{jt} for every job j and every time period $(t, t+1]$ representing the fraction of this period that is dedicated to the processing of job j.

$$\text{minimize} \quad \sum_{j \in J} w_j C_j$$

$$(LP) \qquad \text{subject to} \quad \sum_{t=r_j}^{T} y_{jt} = p_j \quad \text{for all } j \in J \tag{1}$$

$$\sum_{j \in J} y_{jt} \leqslant 1 \quad \text{for } t = 0, \ldots, T \tag{2}$$

$$\frac{1}{p_j} \sum_{\ell=r_j}^{t} y_{j\ell} \geqslant \frac{1}{p_k} \sum_{\ell=r_k}^{t} y_{k\ell} \quad \text{for all } j \prec' k \text{ and } t = 0, \ldots, T \tag{3}$$

$$C_j = \frac{p_j}{2} + \frac{1}{p_j} \sum_{t=0}^{T} y_{jt} \left(t + \tfrac{1}{2} \right) \quad \text{for all } j \in J \tag{4}$$

$$y_{jt} = 0 \quad \text{for all } j \in J \text{ and } t = 0, \ldots, r_j - 1 \tag{5}$$

$$y_{jt} \geqslant 0 \quad \text{for all } j \in J \text{ and } t = r_j, \ldots, T \tag{6}$$

Equations (1) say that all the fractions of a job, which are processed in accordance with the release dates, must sum up to the whole job. Since the machine can process only one job at a time, the machine capacity constraints (2) must be satisfied. Constraints (3) say that at any point $t + 1$ in time the completed fraction of job k must not be larger than the completed fraction of its predecessor j.

Consider an arbitrary feasible schedule, where job j is being continuously processed between $C_j - p_j$ and C_j on the machine. Then the expression for C_j in (4) corresponds to the real completion time, if we assign the values to the LP variables y_{jt} as defined above, i. e., $y_{jt} = 1$ if j is being processed in the time interval $(t, t+1]$. If j is not being continuously processed but preempted once or several times, the expression for C_j in (4) is a lower bound for the real completion time. Hence, (LP) is a relaxation of the scheduling problem under consideration.

On the other hand, we can interpret every feasible solution y to (LP) in a natural way as a fractional schedule which, by (3), softly respects the precedence constraints: take a linear extension of the precedence constraints and process in each time interval $(t, t+1]$ the occurring jobs j for time y_{jt} each, in this order; see Fig. 1 for an example.

In the following, we always identify a feasible solution to (LP) with a corresponding fractional schedule and vice versa. We mutually use the interpretation that seems more suitable for our purposes.

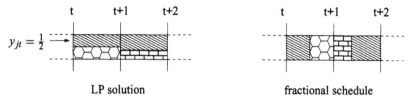

LP solution   fractional schedule

Fig. 1. Interpretation of an LP solution as a fractional schedule and vice versa

Thus, from the LP we get a fractional schedule that satisfies the soft precedence constraints and a lower bound on its total weighted completion time. Of course, this also is a lower bound for preemptive schedules obeying the precedence constraints, and, in turn, of nonpreemptive schedules respecting the precedence constraints. In other words, the LP is a relaxation of $1|r_j, pmtn, prec'|\sum w_j C_j$, $1|r_j, pmtn, prec|\sum w_j C_j$, and $1|r_j, prec|\sum w_j C_j$. The following example shows that it is not better than a 2–relaxation for fractional scheduling (and therefore for the other cases as well) even if all the release dates are 0 and all processing times are 1: let $J = \{1, \ldots, n\}$, $w_j = 0$ for $1 \leqslant j \leqslant n-1$, $w_n = 1$ and $j \prec' n$ for $1 \leqslant j \leqslant n-1$. We get a feasible LP solution if we schedule in every time interval $(t, t+1]$ for $0 \leqslant t \leqslant n-1$ a $\frac{1}{n}$–fraction of each job. This yields the LP completion time $(n+1)/2$ for job n, whereas its optimum completion time is obviously n.

The following lemma highlights the relation between the LP completion time and the completion time in the corresponding fractional schedule for a feasible solution to (LP). The observation in part a) is due to Goemans [Goe97]; an analogous result to part b) was already given in [HSW96, Lemma 2.1] for a somewhat different LP.

Lemma 3. *Consider a feasible solution to* (LP) *and let* C_j^{LP} *be the LP completion time of* j *defined in* (4). *Denote the real completion time in the corresponding fractional schedule by* C_j. *Then the following holds:*

a) $\int_0^1 C_j(\alpha)\, d\alpha \leqslant C_j^{LP}$;

b) $C_j(\alpha) \leqslant \frac{1}{1-\alpha} C_j^{LP}$ *for any constant* $\alpha \in (0, 1)$ *and the given bound is tight.*

Proof. In order to prove part a) we denote by α_t the fraction of job j that is in the fractional schedule completed at time t, for $t = 0, \ldots, T+1$. Thus $(\alpha_t)_{t=0}^{T+1}$ is a monotonically nondecreasing sequence with $\alpha_0 = 0$, $\alpha_{T+1} = 1$ and $C_j(\alpha) \leqslant t+1$ for $\alpha \leqslant \alpha_{t+1}$. We can therefore write

$$\int_0^1 C_j(\alpha)\, d\alpha = \sum_{t=0}^{T} \int_{\alpha_t}^{\alpha_{t+1}} C_j(\alpha)\, d\alpha \leqslant \sum_{t=0}^{T} (\alpha_{t+1} - \alpha_t)(t+1)$$

$$= \frac{1}{p_j} \sum_{t=0}^{T} y_{jt}(t+1) = \tfrac{1}{2} + \frac{1}{p_j} \sum_{t=0}^{T} y_{jt}(t+\tfrac{1}{2}) \leqslant C_j^{LP} \ .$$

The last equality follows from (1). As a consequence of part a) we get for $0 < \alpha \leqslant 1$

$$(1-\alpha)C_j(\alpha) \leqslant \int_\alpha^1 C_j(x)\, dx \leqslant \int_0^1 C_j(x)\, dx \leqslant C_j^{LP} \ .$$

In order to prove the tightness of this bound, we consider a job j with $p_j = 1$, $r_j = 0$ and an LP solution satisfying $y_{j0} = \alpha - \varepsilon$ and $y_{jT} = 1 - \alpha + \varepsilon$, where $\varepsilon > 0$. This yields $C_j^{LP} = 1 + T(1 - \alpha + \varepsilon)$ and $C_j(\alpha) \geqslant T$. Thus, for ε arbitrarily small and T arbitrarily large, the given bound gets tight. □

As a consequence of Lemma 3 part b) we know that the value of the fractional schedule given by an optimum LP solution can be arbitrarily bad compared to the LP value. Given an arbitrary LP solution and some $\beta \geqslant 1$, the following algorithm computes a new schedule such that all the completion times of jobs are within a constant factor of their LP completion times.

Algorithm SLOW–MOTION
Input: fractional schedule S obeying release dates and soft precedence constraints, and a parameter $\beta \geqslant 1$.
Output: fractional schedule that obeys release dates and soft precedence constraints.

slow motion   Consider the given schedule S as a movie and display it β times slower than in real time. Mathematically spoken, we map every point t in time onto $\beta \cdot t$. This defines in a natural way a new schedule S' where job j is being processed for $\beta \cdot p_j$ time units.

cut   Let t_j be the earliest point in time when job j has been processed for p_j time units in S'. Convert S' to a feasible schedule S'' by leaving the machine idle whenever it processed job j after t_j.

The output of Algorithm SLOW–MOTION still allows fractional preemption and only obeys the soft precedence constraints. But we can overcome these drawbacks if we use Algorithm SOFT–TO–STRONG as a postprocessing step. Figure 2 shows the action of Algorithm SLOW–MOTION and the postprocessing with Algorithm SOFT–TO–STRONG for an example: we are given three jobs $J = \{1, 2, 3\}$ with $p_1 = p_2 = p_3 = 2$, $r_1 = 0$, $r_2 = r_3 = 1$ and $2 \prec' 3$. The parameter β is set to $\frac{3}{2}$ in this example.

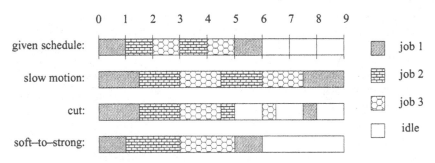

Fig. 2. The two steps in Algorithm SLOW–MOTION together with postprocessing by Algorithm SOFT–TO–STRONG.

Lemma 4. *Algorithm* SLOW–MOTION *with input* $\beta \geqslant 1$ *computes a feasible fractional schedule. The completion time of job j equals* $\beta \cdot C_j(\frac{1}{\beta})$ *where C_j is the completion time of job j in the given schedule.*

Proof. Let C_j^β denote the completion time of job j in the schedule constructed by Algorithm SLOW–MOTION. By construction $C_j^\beta(\alpha) = \beta \cdot C_j(\alpha/\beta)$ for $0 < \alpha \leqslant 1$. In particular $C_j^\beta(0^+) = \beta \cdot C_j(0^+) \geqslant \beta \cdot r_j \geqslant r_j$ and $C_j^\beta = \beta \cdot C_j(\frac{1}{\beta})$. Moreover, since S respects the soft precedence constraints we get

$$C_j^\beta(\alpha) = \beta \cdot C_j(\alpha/\beta) \leqslant \beta \cdot C_k(\alpha/\beta) = C_k^\beta(\alpha)$$

for $j \prec' k$ and $0 < \alpha \leqslant 1$. Thus the constructed schedule is feasible. $\square$

We would like to mention that Lemma 4 is related to Theorem 2.1 in [HSW96]. As an easy corollary of Lemma 1, Lemma 3 b), and Lemma 4, Algorithm SLOW–MOTION followed by Algorithm SOFT–TO–STRONG has a performance guarantee of $\beta^2/(\beta - 1)$ for the problem $1 \mid r_j, pmtn, prec' \mid \sum w_j C_j$. An optimal choice seems to be $\beta = 2$ yielding the performance guarantee 4. But one can in fact do better by choosing β randomly:

Theorem 5. *If one invokes Algorithm* SLOW–MOTION *with input β such that $1/\beta$ is randomly drawn from $(0, 1]$ with density function $f(x) = 2x$, then the expected completion time of a job j in the resulting schedule is bounded from above by twice its LP completion time C_j^{LP}.*

Proof. By Lemma 3 a) and Lemma 4 the expected completion time of j is

$$\int_0^1 f(x) \frac{1}{x} C_j(x) \, dx = 2 \int_0^1 C_j(x) \, dx \leqslant 2 \cdot C_j^{LP} \ .$$
$\square$

Since the value of an optimum LP-solution is a lower bound on the value of any feasible schedule, this randomized algorithm has performance guarantee 2 if we start with an optimum LP-solution. Using Lemma 1 and Corollary 2 we have thus found randomized 2–approximation algorithms for the following problems:

$$1 \mid r_j, pmtn, prec' \mid \sum w_j C_j, \quad 1 \mid r_j, pmtn, prec \mid \sum w_j C_j, \quad \text{and} \quad 1 \mid prec \mid \sum w_j C_j$$

Moreover, we have shown that (LP) is a 2–relaxation for these problems and that this is best possible for (LP). For both problems $1 \mid r_j, pmtn, prec \mid \sum w_j C_j$ and $1 \mid prec \mid \sum w_j C_j$ 2–approximation algorithms based on different LP relaxations and without any usage of randomness were known before; see [HSSW96] for both algorithms. In addition, Goemans [Goe96] applied the very same technique as in Theorem 5 to improve the 4–approximation algorithm of Hall, Shmoys, and Wein [HSW96] for $1 \mid prec \mid \sum w_j C_j$ to a 2–approximation algorithm.

Notice that we cannot immediately solve (LP) in polynomial time since there are exponentially many variables y_{jt}. Thus, up to now, we have only given pseudo-polynomial approximation algorithms. However, we can overcome this drawback by introducing new variables which are not associated with exponentially many time intervals of length 1, but rather with a polynomial number of intervals of geometrically increasing size. This idea was introduced earlier by Hall, Shmoys, and Wein [HSW96] and since then used in several settings, e. g., in [SS97].

However, in order to get polynomial-time approximation algorithms in this way, we have to pay for with a slightly worse performance guarantee. For any constant $\varepsilon > 0$ we get polynomial-time approximation algorithms with performance guarantee $2 + \varepsilon$ for the scheduling problems listed above.

3 An e–Approximation Algorithm for the General Constrained One–Machine Scheduling Problem

As mentioned earlier, the randomized conversion technique of [CMNS97] transforms any preemptive schedule into a nonpreemptive one such that the expected completion time of job j in the nonpreemptive schedule is at most a factor of $\frac{e}{e-1}$ of its completion time in the preemptive schedule. In addition, it is easy to see that, if the preemptive schedule obeys given precedence constraints the same holds for the produced nonpreemptive schedule. Hence, if we first use SLOW–MOTION followed by SOFT–TO–STRONG to get a 2–approximation for $1 | r_j, pmtn, prec | \sum w_j C_j$ and then invoke the conversion algorithm of Chekuri et al., we get a $\frac{2e}{e-1}$–approximation algorithm for $1 | r_j, prec | \sum w_j C_j$. Unfortunately, this does not immediately lead to a better performance guarantee than the 3–approximation algorithm known before [Sch96]. However, when we combine both algorithms in a somewhat more intricate way we get a considerably improved performance guarantee. The key idea is that instead of using the algorithms independently we make the choice of the random variable in the algorithm of Chekuri et al. dependent on the choice of β in Algorithm SLOW–MOTION.

Consider the following algorithm that makes use of density functions f and g_β which are subsequently defined.

Algorithm: e–APPROXIMATION
Input: Instance of $1 | r_j, prec | \sum w_j C_j$.
Output: Feasible schedule.
1) Compute an optimum solution to (LP). Call the corresponding fractional schedule S.
2) Draw $\frac{1}{\beta}$ randomly from $(0, 1]$ with density function f.
3) Let S' be the fractional schedule produced by SLOW–MOTION with input (S, β).
4) Draw α randomly from $(0, 1]$ with density function g_β.
5) Construct a nonpreemptive schedule by scheduling the jobs as early as possible in nondecreasing order of $C'_j(\alpha)$.

Here, C'_j is the completion time of job j in the schedule S'. Recall that SLOW–MOTION guarantees that $C'_j(\alpha) \leqslant C'_k(\alpha)$ for $j \prec k$ and any $\alpha \in (0, 1]$. Consequently the schedule produced by Algorithm e–APPROXIMATION is indeed feasible.

Lemma 6. *Let* $\beta \geqslant 1$ *be a constant and* $g_\beta(x) = \frac{1/\beta}{e^{1/\beta} - 1} \cdot e^{x/\beta}$, *for* $x \in (0, 1]$. *Then, for each job* $j \in J$, *its expected completion time* $E_\beta(C_j)$ *in the schedule constructed by* e–APPROXIMATION *is at most* $\frac{1}{\beta} \cdot \frac{e^{1/\beta}}{e^{1/\beta} - 1} \cdot C'_j$.

Proof. (*Sketch.*) Our analysis almost follows the analysis of [CMNS97] for their conversion technique with the small, but crucial exception that we utilize the structure of the fractional schedule S' produced by Algorithm SLOW–MOTION. In fact, as in [CMNS97] and for the purpose of a more accessible analysis, we do not analyze the schedule produced by e–APPROXIMATION but rather a more structured one which is, however, not better than the output of e–APPROXIMATION. This schedule is obtained by replacing Step 5 with the following procedure:

5') Take the fractional schedule S' produced by Algorithm SLOW–MOTION. Consider the jobs $j \in J$ in nonincreasing order of $C'_j(\alpha)$ and iteratively change the current preemptive schedule by applying the following steps:
   i) remove the $\alpha \cdot p_j$ units of job j that are processed before $C'_j(\alpha)$ from the machine and leave it idle within the corresponding time intervals; we say that this idle time is caused by job j;
   ii) postpone the whole processing that is done later than $C'_j(\alpha)$ by p_j;
   iii) remove the remaining $(1 - \alpha)$–fraction of job j from the machine and shrink the corresponding time intervals;
   iv) process job j in the released time interval $(C'_j(\alpha), C'_j(\alpha) + p_j]$.

Consider an arbitrary but fixed $\alpha \in (0, 1]$ and a job j. We denote with C^α_j its completion time in this new schedule. Let K_α be the set of jobs that complete an α–fraction in S' before $C'_j(\alpha)$, i. e., $K_\alpha = \{k \in J : C'_k(\alpha) \leqslant C'_j(\alpha)\}$. Moreover, for a job $k \notin K$ let $\eta_k(\alpha)$ be the fraction of k that is completed in S' by time $C'_j(\alpha)$. Then, because of Step 5', it is not too difficult to see that

$$C^\alpha_j \leqslant T_j + (1+\alpha) \sum_{k \in K_\alpha} p_k + \sum_{k \notin K_\alpha} \eta_k(\alpha) \cdot p_k$$

where T_j is the total idle time in the schedule S' before j completes. The factors α and $\eta_k(\alpha)$ purely result from the idle time caused by the respective jobs. This is the moment we can improve the analysis by exploiting the structure of S'. First, observe that the necessity of idle time in the schedule is only caused by release dates. Second, because of the slow motion, no job j starts before time $\beta \cdot r_j$ in S'. Consequently, we may further modify the output of Step 5' by reshrinking the idle time caused by job j by a factor of β, for all $j \in J$, without violating the feasibility. Therefore we get

$$C^\alpha_j \leqslant T_j + (1+\frac{\alpha}{\beta}) \sum_{k \in K_\alpha} p_k + \sum_{k \notin K_\alpha} \frac{\eta_k(\alpha)}{\beta} \cdot p_k \ .$$

Hence,

$$E_\beta(C_j) \leqslant \int_0^1 \left(T_j + (1+\frac{\alpha}{\beta}) \sum_{k \in K_\alpha} p_k + \sum_{k \notin K_\alpha} \frac{\eta_k(\alpha)}{\beta} \cdot p_k \right) g_\beta(\alpha) \, d\alpha \tag{7}$$

and a similar argument as the one given by Chekuri et al. [CMNS97] gives the desired bound. □

Lemma 6 together with Lemma 4 yields the following upper bound on the expected completion time of any job j in the schedule produced by Algorithm e–APPROXIMATION:

$$E(C_j) \leqslant \int_0^1 f(x) \cdot \frac{e^x}{e^x - 1} \cdot C_j^S(x) \, dx \; ,$$

where C_j^S is the completion time of job j in schedule S and $\frac{1}{\beta}$ is replaced by x. An optimal choice of the density function f is $f(x) = e \cdot (1 - e^{-x})$. Hence,

$$E(C_j) \leqslant e \int_0^1 C_j^S(x) \, dx \leqslant e \cdot C_j^{LP}$$

by Lemma 3 a). This eventually gives the following result.

Theorem 7. *Algorithm e–APPROXIMATION is an e–approximation algorithm for single machine scheduling subject to release dates and precedence constraints so as to minimize the average weighted completion time.*

To summarize where the improvement on $\frac{2e}{e-1}$ originates from, observe that the bound (7) is the better the larger β is chosen since the idle time caused by jobs is shrunken by the factor β. On the other hand, Lemma 4 and Lemma 3 b) show that the bound on the completion time of any specific job increases as $\beta \geqslant 2$ increases. It is the balancing of these two effects that leads to the better approximation.

As a consequence of Lemma 4 we know that $C_j'(\alpha) = \beta \cdot C_j^S(\alpha/\beta)$. Hence, Algorithm e–APPROXIMATION nonpreemptively schedules the jobs as early as possible in nondecreasing order of $C_j^S(\gamma)$ where $\gamma = \alpha/\beta$ is randomly drawn from $(0,1]$ with a certain density function implicitly defined by f and g_β.

Finally, observe that (LP) therefore is an e–relaxation of $1|r_j, prec|\sum w_j C_j$. Due to the huge number of variables in this LP relaxation, e–APPROXIMATION also is only a pseudo-polynomial algorithm. Again, however, we may replace this LP with a similar one in interval-indexed variables and this results in a polynomial-time $(e + \varepsilon)$–approximation algorithm for the general constrained one-machine scheduling problem $1|r_j, prec|\sum w_j C_j$.

4   List Scheduling with Random Assignments to Parallel Machines

We consider the scheduling problem $R||\sum w_j C_j$. That is, we are given m machines and the processing time of each job $j \in J$ may depend on the machine i it is processed on; it is therefore denoted by p_{ij}. Each job must be processed on one of the machines, and may be assigned to any of them. In contrast to [SS97], we use a tighter LP relaxation in order to get a lower bound for the problem which we then use to give an improved approximation algorithm. Independently, this improvement has also been derived by Fabián A. Chudak (communicated to us by David B. Shmoys, March 1997) after reading a preliminary version of [SS97] which contained a 2–approximation algorithm for $R|r_{ij}|\sum w_j C_j$.

Let $T = \sum_{j \in J} \max_i p_{ij} - 1$ and introduce for every job $j \in J$, every machine $i = 1, \dots, m$, and every point $t = 0, \dots, T$ in time a variable y_{ijt} which represents the processing time of job j on machine i within the time interval $(t, t+1]$. Equivalently, one can say that a y_{ijt}/p_{ij}–fraction of job j is being processed on machine i within the time interval $(t, t+1]$. The LP is as follows:

$$\text{minimize} \quad \sum_{j \in J} w_j C_j$$

(LP_R) subject to
$$\sum_{i=1}^{m} \sum_{t=0}^{T} \frac{y_{ijt}}{p_{ij}} = 1 \quad \text{for all } j \in J \tag{8}$$

$$\sum_{j \in J} y_{ijt} \leqslant 1 \quad \text{for } i = 1, \ldots, m \text{ and } t = 0, \ldots, T \tag{9}$$

$$C_j \geqslant \sum_{i=1}^{m} \sum_{t=0}^{T} \left(\frac{y_{ijt}}{p_{ij}} \left(t + \tfrac{1}{2}\right) + \tfrac{1}{2} y_{ijt} \right) \quad \text{for all } j \in J \tag{10}$$

$$C_j \geqslant \sum_{i=1}^{m} \sum_{t=0}^{T} y_{ijt} \quad \text{for all } j \in J \tag{11}$$

$$y_{ijt} \geqslant 0 \quad \text{for } i = 1, \ldots, m, \; j \in J, \; t = 0, \ldots, T \tag{12}$$

Equations (8) say that all the fractions of a job must sum up to the processing requirement of the whole job. Since each machine can process only one job at a time, the machine capacity constraints (9) must be satisfied. Consider an arbitrary feasible schedule, where job j is being continuously processed between $C_j - p_{hj}$ and C_j on machine h. Then the right hand side of (10) corresponds to the real completion time, if we assign the values to the LP variables y_{ijt} as defined above, i.e., $y_{ijt} = 1$ if $i = h$ and $t \in [C_j - p_{hj}, C_j - 1]$, and $y_{ijt} = 0$ otherwise. Moreover, the right hand side of (11) equals the processing time p_{hj} of j on machine h in this case and is therefore a lower bound on the completion time. Hence, (LP_R) is a relaxation of the scheduling problem under consideration.

We can divide the problem to construct a feasible schedule for a given instance into two steps: in the first step we assign each job to a machine it should be processed on; in the second step we decide in which order the jobs are processed on each of the m machines. In fact, we only have to worry about the first step since the second step can be solved optimally by Smith's rule [Smi56]: sequence for each machine i the jobs that were assigned to it in order of nonincreasing w_j/p_{ij}. So it remains to decide which job should be processed on which machine. This is the point where we can make use of an optimum solution to (LP_R). We may interpret the LP solution as follows: for job j a fraction of it given by $f_{ij} = \frac{1}{p_{ij}} \sum_{t=0}^{T} y_{ijt}$ should be processed on machine i. By (8) these fractions some up to 1 over all machines. Since we are not allowed to divide j into fractions but have to process it continuously on one machine, we interpret the f_{ij} as probabilities instead. This yields the following algorithm:

Algorithm RANDOM–ASSIGNMENT
Input: Instance of $R \mid \mid \sum w_j C_j$.
Output: Feasible schedule.
1) Compute an optimum solution to (LP_R).
2) For each job $j \in J$, assign job j to machine i with probability f_{ij}.
3) Schedule on each machine i the jobs that were assigned to it in order of nonincreasing w_j/p_{ij}.

Theorem 8. *Algorithm* RANDOM–ASSIGNMENT *has performance guarantee* $\frac{3}{2}$.

In order to prove Theorem 8 we in fact do not consider Algorithm RANDOM–ASSIGNMENT but a modified version which is slightly more complicated, not better than the original algorithm, but easier to analyze. We replace Steps 2 and 3 by

2') Assign job j to a machine-time pair (i,t) with probability y_{ijt}/p_{ij} and set $t_j := t$.
3') Schedule on each machine i the jobs that were assigned to it in order of nondecreasing t_j.

The random assignment of job j to one of the machines in Step 2' is exactly the same as in the original algorithm. In addition, we create a random order through the random variables t_j, ties are also broken randomly. Of course, this random order cannot be better than the optimum order given by Smith's rule. Thus the modified algorithm cannot be better than the original one and Theorem 8 is a corollary of the following lemma.

Lemma 9. *The expected completion time of job j in the schedule constructed by the modified algorithm is bounded from above by $\frac{3}{2}$ times the optimum LP completion time C_j^{LP}, for all $j \in J$.*

Proof. We first consider the conditional expectation of j's completion time under the assumption that j was assigned to the machine-time pair (i,t). We denote this conditional expectation by $E_{it}(C_j)$ and get

$$E_{it}(C_j) = p_{ij} + \sum_{k \neq j} p_{ik} \cdot \Pr(k \text{ on } i \text{ before } j) = p_{ij} + \sum_{k \neq j} p_{ik} \cdot \left(\sum_{\ell=0}^{t-1} \frac{y_{ik\ell}}{p_{ik}} + \frac{1}{2} \cdot \frac{y_{ikt}}{p_{ik}} \right)$$

$$= p_{ij} + \sum_{\ell=0}^{t-1} \sum_{k \neq j} y_{ik\ell} + \frac{1}{2} \sum_{k \neq j} y_{ikt} \leq p_{ij} + t + \frac{1}{2} \ .$$

The last inequality follows from the machine capacity constraints (9). Summing over all possible assignments of j to intervals and machines we finally bound the expectation of j's completion time as follows:

$$E(C_j) = \sum_{i=1}^{m} \sum_{t=0}^{T} \frac{y_{ijt}}{p_{ij}} \cdot E_{it}(C_j) \leq \sum_{i=1}^{m} \sum_{t=0}^{T} \left(\frac{y_{ijt}}{p_{ij}} (t + \frac{1}{2}) + y_{ijt} \right)$$

$$= \sum_{i=1}^{m} \sum_{t=0}^{T} \left(\frac{y_{ijt}}{p_{ij}} (t + \frac{1}{2}) + \frac{1}{2} y_{ijt} \right) + \frac{1}{2} \sum_{i=1}^{m} \sum_{t=0}^{T} y_{ijt} \leq \frac{3}{2} \cdot C_j^{LP} \ .$$

The last inequality follows from (10) and (11). □

Again, we cannot directly solve (LP_R) in polynomial time since there are exponentially many variables y_{ijt}. Thus the running time of Algorithm RANDOM–ASSIGNMENT is only pseudo-polynomial. However, once more we can overcome this difficulty by replacing the time-indexed variables with interval-indexed variables. Hence, we get a polynomial-time approximation algorithm with performance guarantee $\frac{3}{2} + \varepsilon$ for any constant $\varepsilon > 0$.

As a special case of the scheduling problem discussed above we now consider identical parallel machines: the processing time of job j does not depend on the machine it is processed on and it is therefore denoted by p_j. A generalization of Smith's rule to identical parallel machines is the following list scheduling algorithm: sort the jobs in order of nonincreasing w_j/p_j; whenever a machine becomes available, the list is scanned for the first not yet executed job, which is then assigned to the machine. Kawaguchi and Kyan [KK86] have shown that this simple algorithm has performance guarantee $(\sqrt{2}+1)/2$ and that this is best possible for the algorithm. Since their analysis is somewhat complicated we present here a randomized algorithm which is as simple as the one of Kawaguchi and Kyan but much easier to analyze. Moreover, its derandomization by the method of conditional probabilities is precisely the algorithm of Kawaguchi and Kyan.

Algorithm RANDOM–KK
Input: Instance of $P \mid \mid \sum w_j C_j$.
Output: Feasible schedule.
1) Assign each job randomly (with probability $\frac{1}{m}$) to one of the machines.
2) Schedule on each machine i the jobs that were assigned to i in order of nonincreasing w_j/p_j.

Theorem 10.
a) *Algorithm* RANDOM–KK *has performance guarantee* $\frac{3}{2}$. *Moreover, there exist instances for which this bound is asymptotically tight.*
b) *Derandomization by the method of conditional probabilities precisely gives the algorithm of Kawaguchi and Kyan.*

Proof. Because of Theorem 8 and the construction of Algorithm RANDOM–ASSIGNMENT it suffices to show that there exists an optimum solution to (LP_R) such that each of the m machines processes an $\frac{1}{m}$ fraction of every job. This can be easily seen in the following way: take an optimum solution y to (LP_R) and construct a new solution y' by setting $y'_{ijt} = \frac{1}{m} \sum_{h=1}^{m} y_{hjt}$ for each i, j and t. This is obviously a feasible solution to (LP_R) with the same objective value and therefore optimal. If we choose this solution in Step 1 of Algorithm RANDOM–ASSIGNMENT, it obviously does the same as Algorithm RANDOM–KK.

In order to show that the performance guarantee $\frac{3}{2}$ is best possible, we consider instances with m identical parallel machines and m jobs of unit length and weight. We get an optimal schedule with value m by assigning one job to each machine. On the other hand we can show that the expected completion time of a job in the schedule constructed by Algorithm RANDOM–KK is $\frac{3}{2} - \frac{1}{2m}$ which converges to $\frac{3}{2}$ for increasing m. Since the fraction w_j/p_j equals 1 for all jobs, we can w. l. o. g. schedule on each machine the jobs that were assigned to it in a random order. Consider a fixed job j and the machine i it has been assigned to. The probability that a job $k \neq j$ was assigned to the same machine is $\frac{1}{m}$. In this case k is processed before j on the machine with probability $\frac{1}{2}$. We therefore get $E(C_j) = 1 + \sum_{k \neq j} \frac{1}{2m} = \frac{3}{2} - \frac{1}{2m}$.

The derandomization of Algorithm RANDOM–KK by the method of conditional probabilities yields: iteratively consider the jobs in order of nonincreasing w_j/p_j and assign job j to the machine with the smallest load so far. This is just another formulation of Kawaguchi and Kyan's algorithm. The smallest load is bounded from above by the average load $\frac{1}{m}\sum_{k<j}p_k$ where we sum over all jobs k that we considered before j. Thus the completion time of j is bounded by $p_j + \frac{1}{m}\sum_{k<j}p_k$ which exactly equals the expected completion time of j in Algorithm RANDOM–KK. $\qquad\square$

As a consequence of the tightness result in Theorem 10 a) we know that the bound in Theorem 8 is also tight.

References

[CMNS97] C. Chekuri, R. Motwani, B. Natarajan, and C. Stein. Approximation techniques for average completion time scheduling. In *Proceedings of the 8th ACM–SIAM Symposium on Discrete Algorithms*, pages 609 – 618, 1997.

[DW90] M. E. Dyer and L. A. Wolsey. Formulating the single machine sequencing problem with release dates as a mixed integer program. *Discrete Applied Mathematics*, 26:255 – 270, 1990.

[GLLRK79] R. L. Graham, E. L. Lawler, J. K. Lenstra, and A. H. G. Rinnooy Kan. Optimization and approximation in deterministic sequencing and scheduling: A survey. *Annals of Discrete Mathematics*, 5:287 – 326, 1979.

[Goe96] M. X. Goemans, June 1996. Personal communication.

[Goe97] M. X. Goemans. Improved approximation algorithms for scheduling with release dates. In *Proceedings of the 8th ACM–SIAM Symposium on Discrete Algorithms*, pages 591 – 598, 1997.

[HSSW96] L. A. Hall, A. S. Schulz, D. B. Shmoys, and J. Wein. Scheduling to minimize average completion time: Off–line and on–line approximation algorithms. Preprint 516/1996, Department of Mathematics, Technical University of Berlin, Berlin, Germany, 1996. To appear in *Mathematics of Operations Research*.

[HSW96] L. A. Hall, D. B. Shmoys, and J. Wein. Scheduling to minimize average completion time: Off–line and on–line algorithms. In *Proceedings of the 7th ACM–SIAM Symposium on Discrete Algorithms*, pages 142 – 151, 1996.

[KK86] T. Kawaguchi and S. Kyan. Worst case bound of an LRF schedule for the mean weighted flow–time problem. *SIAM Journal on Computing*, 15:1119 – 1129, 1986.

[Law78] E. L. Lawler. Sequencing jobs to minimize total weighted completion time subject to precedence constraints. *Annals of Discrete Mathematics*, 2:75 – 90, 1978.

[PSW95] C. Phillips, C. Stein, and J. Wein. Scheduling jobs that arrive over time. In *Proceedings of the Fourth Workshop on Algorithms and Data Structures*, number 955 in Lecture Notes in Computer Science, pages 86 – 97. Springer, Berlin, 1995.

[Sch96] A. S. Schulz. Scheduling to minimize total weighted completion time: Performance guarantees of LP–based heuristics and lower bounds. In W. H. Cunningham, S. T. McCormick, and M. Queyranne, editors, *Integer Programming and Combinatorial Optimization*, number 1084 in Lecture Notes in Computer Science, pages 301 – 315. Springer, Berlin, 1996. Proceedings of the 5th International IPCO Conference.

[Smi56] W. E. Smith. Various optimizers for single–stage production. *Naval Research and Logistics Quarterly*, 3:59 – 66, 1956.

[SS97] A. S. Schulz and M. Skutella. Scheduling–LPs bear probabilities: Randomized approximations for min–sum criteria. Preprint 533/1996, Department of Mathematics, Technical University of Berlin, Berlin, Germany, November 1996; revised March 1997. To appear in Springer Lecture Notes in Computer Science, Proceedings of the 5th Annual European Symposium on Algorithms (ESA'97).

'Go with the Winners' Generators with Applications to Molecular Modeling[*]

Marcus Peinado and Thomas Lengauer

Institute for Algorithms and Scientific Computing
German National Research Center for Information Technology (GMD)
53754 Sankt Augustin, Germany
Email: {mpe, lengauer}@cartan.gmd.de

Abstract. A generation problem is the problem of generating an element of a (usually exponentially large) set under a given distribution. We develop a method for the design of generation algorithms which is based on the 'go with the winners' algorithm of Aldous and Vazirani [AV94]. We apply the scheme to two concrete problems from computational chemistry: the generation of models of amorphous solids and of certain kinds of polymers.

1 Introduction

The generation problem for a given finite set R and a given probability distribution π on R is the task of outputting a random element of R under distribution π. In contrast, the construction problem corresponds to the potentially easier task of producing an arbitrary element of R. In general, a generation problem has an input x which specifies the set R_x from which elements are to be generated and a corresponding distribution π_x. Typically, R_x is exponentially large in the size of the input x, and a solution to the generation problem represents a way of obtaining a 'typical' representative of the set. General treatments of generation problems and their relation to counting problems are contained in [JVV86, SJ89].

So far, the most successful general algorithmic tool for the design of generation algorithms has been the theory of rapidly mixing Markov chains (e.g. [Sin93]). Abstractly, the algorithm designer defines an ergodic Markov chain with stationary distribution π_x on state space R_x by choosing appropriate transition probabilities between pairs of elements of R_x. The generation algorithm simulates the Markov chain for a given number t of steps and outputs its state at time t. In certain cases, it can be shown that the output distribution is close to π_x, even if t is only poly-logarithmic in the size of R_x.

In this paper, we consider a different generation scheme which is based on an extension of the 'go with the winners' algorithm of Aldous and Vazirani [AV94]. Assume the elements of R_x can be viewed as compositions of $l_x \in \mathbb{N}$ elements

---

[*] Supported in part by DFG Grant SFB408. Part of this research was done during the first author's stay at the International Computer Science Institute, Berkeley.

of a smaller set Σ. For example, R_x might be some set of strings of length l_x over an alphabet Σ. Any element of R_x can be constructed by concatenating l_x elements of Σ. The following method for constructing elements of R_x – even though unlikely to succeed in general – is instructive: *Let Y be the empty string and proceed in l_x steps. In each step, select a random element of Σ (from a distribution to be described below) and append it to Y. If it becomes clear that Y cannot be extended into any element of R_x (i.e. that no element of R_x has prefix Y), abort.*

An algorithm of this kind can be described equivalently as a walk from the root of a tree to a node at depth l_x. We will use this analogy throughout the paper. The tree has nodes for all $y \in R_x$ and for all prefixes of elements of R_x. The root of the tree is the empty string. There is an edge from node w to node v if v can be obtained by appending an element of Σ to w (i.e. if $\exists a \in \Sigma : v = wa$). Thus, the nodes at depth i correspond to strings of length i. R_x is exactly the set of nodes at depth l_x. Thus, for every $y \in R_x$, there is a path of length l_x from the root to y. The i-th node in the path is the length i prefix of y.

So far, the tree contains a node *if and only if* it corresponds to a prefix of some element of R_x. We will be less restrictive and broaden the class of admissible trees. We admit any tree which satisfies the following conditions: (a) The set of nodes at depth l_x is R_x. (b) If v is a node corresponding to a string of length $l > 0$ then the length $l-1$ prefix w of v is also a node. There is an edge from w to v. (c) There are no other edges. We call such a tree an R_x-search-tree. In addition to R_x and all its prefixes, the tree may contain nodes at depths less than l_x which do not have descendants at depth l_x. The simple construction algorithm can now be reformulated as Alg. 0 of [AV94]: *"Start at the root, repeatedly choose a child at random (distribution to be determined) until reaching a leaf, then stop."* The algorithm can be visualized as a particle which starts at the root and moves down the tree until it reaches a leaf.

It is not necessary to store the entire tree. An efficiently computable predicate P on Σ^*, such that $P(y)$ is true if and only if string y is a node of the tree, is sufficient. P can be viewed as a pruning procedure. The children of any given node w can be computed by evaluating $P(wa)$ for all $a \in \Sigma$. This is sufficient to implement Alg. 0.

Two problems of the simple algorithm are evident: 1. It must be insured that, conditional on reaching depth l_x, the probability of finding any particular solution $y \in R_x$ coincides with $\pi_x(y)$. 2. Depending on the structure of the tree, the algorithm may be extremely unlikely to reach depth l_x at all. While we cannot hope to reach depth l_x efficiently in every possible tree, the simple algorithm fails even for many trees whose depth l_x nodes can be found trivially by other methods.

In order to resolve the first problem, we assign a weight (or probability) to each edge. We will show how to choose the weights such that a particle which moves down the tree, choosing at each step the next node with probability proportional to the weight of the corresponding edge, reaches node $y \in R_x$ at depth l_x with probability proportional to $\pi_x(y)$.

Concerning the second problem, we replace Alg. 0 by the 'go with the winners' algorithm of Aldous and Vazirani [AV94]. 'Go with the winners' was designed as an optimization algorithm on trees. The goal is to find a node at a given depth l_x. Aldous and Vazirani derive a bound which relates the running time and the success probability of 'go with the winners', and which shows that 'go with the winners' can achieve an exponential improvement over Alg. 0. Still, bounding the running time of 'go with the winners' remains the main task when trying to apply it to any particular application.

Even provided that 'go with the winners' finds nodes at depth l_x, it will, in general, not do so with distribution π_x. We make a small extension to 'go with the winners' and show in the subsequent analysis that the output distribution of the resulting algorithm is close to π_x by bounding the variation distance between the two distributions. The bound depends on the running time allocated to 'go with the winners'. Overall, a constant factor increase over the running time of basic 'go with the winners' is sufficient to make the bound on the variation distance small. This reduces the problem of bounding the running time of our generator to the problem of bounding the running time of basic 'go with the winners'. By the analysis of Aldous and Vazirani [AV94], the running time of 'go with the winners' can be bounded by estimating a certain parameter κ of the tree. This parameter plays a similar role for our generation scheme as does the second-largest eigenvalue of the transition matrix or the conductance for generators based on rapidly mixing Markov chains: When applying the 'go with the winners' generator to a concrete generation problem, the main task is to bound κ. If it can be shown that κ is bounded by a polynomial, the 'go with the winners' generator will run in polynomial time.

Section 2 describes the 'go with the winners' generator, and Sect. 3 describes how this generator can be used to solve the following two generation problems:

1.1 Applications in Molecular Modeling

Many atomic structures can be modeled on the basis of graphs embedded in 3-space. Nodes in the graphs represent atoms (or groups of atoms), edges represent chemical bonds. The need for computer models arises in the chemical analysis of the atomic structure of certain substances. Usually, only partial information about the atomic structure can be obtained by means of measurements. This information might include the distributions of the bond lengths and bond angles. Other pieces of information, such as the atom coordinates themselves, are hard or impossible to obtain experimentally. A common approach among chemists is to produce *representative* computer models of the substance from which the information of interest can be easily derived. For ease of exposition, we consider the following problems only in two dimensions. The generalization to higher dimensions is straightforward.

Polymers and self-avoiding walks: Given a graph G, the task is to generate simple paths of length d (i.e. paths of length d containing no vertex more than once). In practice, interest focuses on G being the d-dimensional square lattice

($d \in \{2,3\}$). These graphs are used as simple discrete models of d-space. It is not known if counting the number of self-avoiding walks of length n on a 2-dimensional square lattice is #P-complete [Wel93].

Efficient solutions for this problem are of considerable importance in the study of the statistical properties of polymers. Several algorithms [RS94, BS85, Gra96] which generate models of polymers (or self-avoiding walks) are based on a tree structure. Randall and Sinclair [RS94] describe an algorithm for the uniform generation of self-avoiding walks whose polynomial time bound depends on a plausible combinatorial hypothesis. The algorithm is based on the simulation of a certain Markov chain. Our analysis implies a similar result for a 'go with the winners' based generator.

Amorphous Solids: This problem has its background in the chemistry of inorganic amorphous solids (covalent glasses) [Ell90]. Such glasses can be modeled as pairs (G, f), where $G = (V, E)$ is a graph whose vertices are mapped into the $n \times n$ square ($n \times n \times n$ cube in three dimensions) by the embedding function $f : V \mapsto \mathbb{R}^2$ such that the following geometric and degree constraints are satisfied (cf. Fig. 1):

1. The length of any edge lies between given minimum and maximum values $d_{\min}$ and $d_{\max}$, where the length of an edge $\{v, w\} \in E$ is defined as the Euclidean distance between $f(v)$ and $f(w)$.
2. The angle between any two edges having one vertex in common lies between a given minimum value $\alpha_{\min}$ and a given maximum value $\alpha_{\max}$.
3. The degree of each vertex is fixed.

The constraints may be modified near the boundary of the square depending on the *boundary conditions*. The input parameter is n, the size of the structure (or the square containing it). In order to make R_n finite, we restrict the range of the embedding function f: the coordinates assigned to the vertices of G must lie on a discrete grid. The distance between neighboring grid points is a small constant fraction of $d_{\max}$. Let $\mathcal{H}(n)$ be the set of all structures which satisfy the constraints.

The problem has both graph theoretic and geometric aspects. If the geometric constraints (edge lengths and angles) are ignored, the problem reduces to randomly generating graphs with a prescribed degree sequence. This problem can be solved in polynomial time by the algorithm of Goldberg and Jerrum [GJ97]. On the other hand, the problem is loosely related to the GRAPH EMBEDDING problem [Hen88, Yem79, Sax79] if the geometric constraints are present but the random generation aspect is ignored.

2 'Go with the Winners' Generators

In this section, we construct 'go with the winners' based generators in a general setting. We proceed in three steps. In this subsection, we show how a given generation problem can be reduced to the problem of finding deep nodes in a

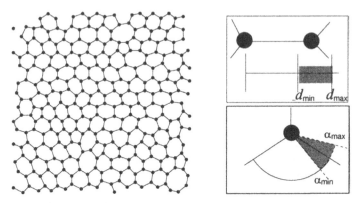

Fig. 1. left: an example of an amorphous structure with vertex degree 3 in two dimensions; right: the bond length and bond angle constraints

certain tree. In Sect. 2.1, we describe the 'go with the winners' algorithm which was designed to find deep nodes in trees. Finally, in Sect. 2.2, we modify 'go with the winners' and show that the output distribution of the resulting algorithm is close to the target distribution.

Following Jerrum et al. [JVV86], we view a uniform generation problem as being given by a relation R – decidable in polynomial time – between problem instances and the set of feasible solutions. Let Σ be a finite alphabet in which we encode problem instances and feasible solutions, and let $R \subseteq \Sigma^* \times \Sigma^*$ be a binary relation between problem instances and feasible solutions such that $(x, y) \in R$ if and only if y is a feasible solution of problem instance x. Given a problem instance x, let $R_x = \{y : (x, y) \in R\}$ be the set of solutions of x. We will assume that all strings in R_x have the same length l_x. This assumption simplifies the exposition considerably. However, it is not fundamental and could, in principle, be removed. For a weighted generation problem it is necessary to specify the output distribution π_x on R_x for each input x.

Definition 1. Consider a generation problem $R \subseteq \Sigma^* \times \Sigma^*$, and a function $wt : \Sigma \to \mathbb{R}$ which assigns a weight to each symbol of Σ. Given a problem instance x, we call a distribution π_x on R_x *wt-multiplicative* if

$$\pi_x(y) = \mathcal{Z}_x^{-1} \prod_{i=1}^{l_x} wt(y_i) \quad \text{for all} \quad y \in R_x \ ,$$

where y_i is the length i prefix of y and $\mathcal{Z}_x = \sum_{y \in R_x} \prod_{i=1}^{l_x} wt(y_i)$ is the normalization factor. We call a sequence of distributions $(\pi_x)_x$ on $(R_x)_x$ *wt-multiplicative* if for all x, π_x is the wt-multiplicative distribution on R_x.

The uniform distribution and the most frequently considered weighted distributions (e.g. [Sin93]) are wt-multiplicative for appropriately chosen Σ and wt.

Let wt be fixed. Any given input x defines a set R_x and a wt-multiplicative distribution π_x. Recall the definition of an R_x-search-tree and Alg. 0 from the introduction. The statement 'at random' in the description of Alg. 0 must be made more precise. Let $p(y)$ be the probability that Alg. 0 visits a given node y. We assign probabilities to the edges of the tree such that for any $y \in R_x$, $p(y)$ is proportional to $\pi_x(y)$. Assign to edge (w, v) the probability

$$p(v|w) = \frac{wt(last(v))}{\sum_{t \in \Sigma} wt(t)} \tag{1}$$

where $last(v) \in \Sigma$ denotes the last symbol of v. Note that the expression in the denominator is constant and that it can be computed in constant time.

The probabilities $p(v|w)$ assigned to the edges leaving any given node w add up to a value q_w which may be smaller (but never larger) than 1. When at w, Alg. 0 should stop and fail with probability $1 - q_w$. This rule can be encoded into the tree itself by appending a leaf node and an edge with weight $1 - q_w$ to any non-leaf node w. Let $V_i \subseteq \Sigma^i$ denote the set of nodes at depth i and let $a(i) = \sum_{v \in V_i} p(v)$ be the probability that the algorithm reaches depth i.

Proposition 2. *Consider a generation problem $R \subseteq \Sigma^* \times \Sigma^*$ and $wt : \Sigma \mapsto \mathbb{R}$. For every instance x, there exists a tree $T = (V, E)$ of depth l_x such that the set of depth l_x nodes is R_x and for each $y \in R_x$*

$$\pi_x(y) = \frac{p(y)}{a(l_x)} \;,$$

where π_x is the wt-multiplicative distribution on R_x.

Proof. T can be any R_x-search-tree with edge weights given by (1) and additional leaf nodes as described in the paragraph following (1). The probability that Alg. 0 reaches any particular solution $y \in R_x$ is

$$p(y) = \prod_{i=0}^{l_x-1} p(y^{(i+1)}|y^{(i)}) = \left(\sum_{t \in \Sigma} wt(t)\right)^{-l_x} \prod_{i=1}^{l_x} wt(y_i)$$

where $y^{(i)}$ denotes the i-th node on the path from the root $y^{(0)}$ to $y = y^{(l_x)}$ and $y_i \in \Sigma$ denotes the i-th alphabet symbol of y. Thus, $\frac{p(y)}{a(l_x)} = \frac{\prod_i wt(y_i)}{\sum_{z \in R_x} \prod_i wt(z_i)} = \pi_x(y)$. □

Thus, conditional on reaching depth l_x, Alg. 0 produces elements of R_x from distribution π_x. The proposition allows us to treat any generation problem with multiplicative target distribution π_x as the problem of generating depth l_x nodes of a tree. The remaining sections will be described within the tree framework.

Example 1: Consider the problem of uniformly generating self-avoiding walks on a two dimensional square lattice, mentioned in the introduction. It can be assumed that one end of the path is the origin. Let $\Sigma = \{N(north), E(east), S(south), W(west)\}$ and let $wt(N) = wt(E) = wt(S) = wt(W) = 1$. Clearly, any path of length d on a 2D lattice graph can be represented as a sequence of d elements of Σ. Alg. 0 corresponds to starting with the path of length 0 and, in each stage, trying to extend the path in any of the four directions with probability $1/\sum_{x\in\{N,E,S,W\}} wt(x) = 1/4$. The predicate P could specify that the run is aborted if the appended grid point is already contained in the path, i.e. if the walk is not self avoiding. This corresponds to a tree in which V_i is the set of self avoiding walks of length i. Clearly, each self-avoiding walk of length d is generated with probability $(1/4)^d$. Thus, conditional on reaching depth d in the tree, the algorithm generates solutions from the uniform distribution. However, the probability that any given run of the algorithm does not abort before reaching depth d, can be very small. For this reason, we consider the 'go with the winners' algorithm.

2.1 The 'Go with the Winners' Algorithm

Given a tree and any two nodes w, v, let $p(w|v)$ be the probability that the particle visits node w given that it is at node v. Let $a(d|v) = \sum_{w\in V_d} p(w|v)$ be the probability that the particle reaches depth $d \in \mathbb{N}$ given that it is at node v. If $a(d)$ is exponentially small, an exponential number of independent runs of Alg. 0 would be needed to obtain a successful run with constant probability. Consider the following alternative to independent runs [AV94, Alg. 2]:

Go with the winners: *Repeat the following procedure, starting at stage 0 with B particles at the root: At stage i there are a random number of particles all at depth i. If all the particles are at leaves then stop. Otherwise, for each particle at a non-leaf, add at that particle's position a random number of particles, this random number having distribution $N(\theta_i^{-1} - 1)$. Particles at leaf nodes are removed. Finally, move each particle from its current position to a child chosen at random.*

B is an input parameter of the algorithm. The θ_i are real constants $0 < \theta_i \leq 1$ (to be specified) and $N(c)$ is an integral valued random variable with expectation c: $\mathbf{Pr}(N(c) = \lfloor c \rfloor) = \lceil c \rceil - c$ and $\mathbf{Pr}(N(c) = \lceil c \rceil) = c - \lfloor c \rfloor$. The goal is to keep the expected number of particles constant over the stages. This is the case if $\theta_i = a(i+1)/a(i)$. The $a(i)$ are generally not directly available. Aldous and Vazirani use the algorithm itself to sample the θ_i recursively:

Sample_θ: *Repeat the following procedure for i from 1 to $d-1$: At the beginning of the i-th iteration, θ_j is already known for $1 \leq j < i$. 'Go with the winners' is run up to stage $i - 1$, leaving all particles at depth i. Then, θ_i is taken as the fraction of particles in non-leaf nodes.*

The main result of [AV94] is that the probability that algorithm *Sample_θ* does not reach the desired depth d is

$$p_{AV} = O\left(B^{-1}\kappa d^4\left(1 + \frac{1}{\beta d}\right)\right) \quad \text{where} \quad \beta = \min_{0\leq i<d}\frac{a(i+1)}{a(i)}, \tag{2}$$

$$\text{and} \quad \kappa = \max_{0\leq i<j\leq d}\kappa_{i,j} = \frac{a(i)}{a^2(j)}\sum_{v\in V_i}p(v)a^2(j|v). \tag{3}$$

Intuitively, κ measures the 'imbalance' of the tree. κ is the main parameter used to characterize the tree and the problem it encodes. Under the assumption that $\beta = \Omega(1/d)$, *(2) means that choosing $B = \theta(\kappa d^4)$ is sufficient to give the algorithm a constant probability of success.*

2.2 The Generator

Assume B is chosen according to (2), such that the probability of reaching depth d is large. Still, the output of 'go with the winners' is not a solution to the generation problem. Every run of 'go with the winners' yields varying numbers of solutions which, in general, are not independent. The dependencies are a simple consequence of the fact that the particles which are added in each stage are copies of existing particles.

Consider any tree, and a given depth d. In this section, we describe how 'go with the winners' can be used to produce independent samples from a distribution which is close to $\pi(v) := p(v)/a(d)$ ($v \in V_d$). Let $U \subseteq V_d$ be an arbitrary subset of V_d and let $p(U) = \sum_{v \in U} p(v)$ and $\pi(U) = \sum_{v \in U} \pi(v)$. Let X_v be the number of particles that reach node v and let $S_i = \sum_{v \in V_i} X_v$ be the number of solutions produced by 'go with the winners' if run up to depth i. The following elementary fact which is implicit in [AV94] is the link between our analysis of Alg. 0 and 'go with the winners'. It states that the expected number of copies of any given solution v is proportional to $\pi(v)$.

Proposition 3. *For all depths i and all $v \in V_i$:* $\mathbf{E}X_v = \frac{p(v)}{a(i)} \mathbf{E}S_i$

Proof. It is straightforward to derive the following (e.g. cf. [AV94, Eq. (8)]):
$$\mathbf{E}X_v = Bp(v)/\prod_{j=1}^{i-1}\theta_j, \quad \mathbf{E}S_i = Ba(i)/\prod_{j=1}^{i-1}\theta_j.$$
□

In light of the possible dependencies of the solutions produced in any single run of 'go-with-the-winners', our generator will not output more than one solution from each run. Note that it is not sufficient to select one particle uniformly, and (unconditionally) output it, as the random variables X_v and S_d are, in general, not independent. We will use the following sampling procedure. The parameter $u > 0$ determines the accuracy of the generator. Its exact value will be determined below.

Algorithm: Gwtw-Gen
Input: desired depth d, $\theta_0, \ldots, \theta_{d-1}$,
1. Run 'go with the winners' up to depth d (producing S_d solutions)
2. With probability $1 - \min\{S_d/u, 1\}$ reject the run and start again at step 1.
3. Select uniformly one of the particles (of the last run) and output it.

Gwtw-gen makes calls to 'go with the winners' (step 1) until a run is accepted in step 2. After that, an output is selected in step 3. The purpose of the acceptance criterion in step 2 is to remove perturbations caused by possible correlations between X_v and S_d. Proposition 4 shows that the output distribution is unperturbed (i.e. exactly π) if runs (step 1) are accepted with probability proportional to S_d. The output distribution of Gwtw-Gen will be perturbed if S_d can exceed u. The event $S_d > u$ can be made less likely by increasing u. At the same time, the average number of runs (i.e. the running time) per output particle increases. The purpose of u is to allow a trade off between accuracy and running time.

The precise nature of this trade off is stated in Thm. 6. The following proposition identifies a special case in which the generator produces exactly the target distribution.

Proposition 4. *Let u be such that $S_d < u$ with probability 1. Let Z be the output of Gwtw-Gen. Then for all $v \in V_d$: $\mathbf{Pr}(Z = v) = \frac{p(v)}{a(d)}$.*

Proof. The proof is a special case of the proof of Lemma 5. Details are omitted due to space restrictions. □

In general, we will not have a sub-exponential (in d) upper bound on $S_d / \mathbf{E}S_d$. In this case, satisfying the precondition $\mathbf{Pr}(S_d < u) = 1$ of Prop. 4 implies choosing u such that it exceeds $\mathbf{E}S_d$ by an exponential factor. By Markov's inequality, this entails that only an exponentially small fraction of the 'go with the winners' runs (step 1) is accepted in step 2. Thus, we will be forced to choose u such that $\mathbf{Pr}(S_d < u) < 1$. In this case, Gwtw-Gen cannot be guaranteed to produce the *exact* target distribution π. Instead, we derive a bound on the variation distance between π and the output distribution of Gwtw-Gen.

We state a basic definition needed in the rest of this section. Let $(\Omega, \mathbf{Pr})$ be a finite probability space. Let X be a random variable with a finite range $\{x_1, \dots, x_k\}$ ($k \in \mathbb{N}$). Then X defines a *partition* of Ω into $(E_1, \dots, E_k)$, where E_i is the event $\{X = x_i\}$ ($1 \leq i \leq k$). Let $\mathbb{1}_E$ denote the indicator variable of a given event E. The *conditional probability of an event A with respect to the random variable X* is defined as $\mathbf{Pr}(A|X) = \sum_{i=1}^{k} \mathbf{Pr}(A|E_i)\mathbb{1}_{E_i}$. Note that $\mathbf{Pr}(A|X)$ is itself a random variable and that $\mathbf{E}\,\mathbf{Pr}(A|X) = \mathbf{Pr}(A)$. The generalization to more than one random variable is obvious. More details can be found in standard texts on probability [Shi84].

Consider any single call to 'go with the winners' by Gwtw-Gen and define the random variable Y as follows: If the run is rejected in step 2, let $Y = \perp$. Otherwise, let Y be the particle selected in step 3.

Lemma 5. *Let $u \geq 4\,\mathbf{E}S_d$. For any $U \subseteq V_d$:*

$$\frac{\mathbf{E}S_d}{u} - \frac{4\,\mathbf{var}(S_d)}{u^2} \leq \mathbf{Pr}(Y \neq \perp) \leq \frac{\mathbf{E}S_d}{u} \tag{4}$$

$$\frac{\mathbf{E}S_d}{u}\frac{p(U)}{a(d)} - \frac{4\,\mathbf{var}(S_d)}{u^2} \leq \mathbf{Pr}(Y \in U) \leq \frac{\mathbf{E}S_d}{u}\frac{p(U)}{a(d)} \tag{5}$$

Proof. By step 2 of Gwtw-Gen

$$\mathbf{Pr}(Y \neq \perp | S_d) = \min\left\{\frac{S_d}{u}, 1\right\} = \frac{S_d}{u} - \frac{S_d - u}{u}\mathbb{1}_{S_d > u} \tag{6}$$

The upper bound of (4) is obvious from this. Concerning the lower bound, note:

$$\mathbf{Pr}(Y \neq \perp) = \mathbf{E}\,\mathbf{Pr}(Y \neq \perp | S_d) = \frac{\mathbf{E}S_d}{u} - \frac{\mathbf{E}((S_d - u)\mathbb{1}_{S_d > u})}{u}$$

$$\geq \frac{\mathbf{E}S_d}{u} - \frac{4\,\mathbf{var}(S_n)}{u^2}$$

It remains to justify the last step. Let $k_i = 2^i u$ ($i \in \mathbb{N}$). Using Chebyshev's inequality, we obtain

$$\mathbf{E}((S_d - u)\mathbb{1}_{S_d > u}) = \sum_{j \geq u} \mathbf{Pr}(S_d > j) \leq \sum_{i=0}^{\infty} k_i \, \mathbf{Pr}(S_d > k_i)$$

$$\leq \sum_{i=0}^{\infty} k_i \frac{\text{var} S_d}{(k_i - \mathbf{E} S_d)^2} = \frac{\text{var} S_d}{u} \sum_{i=0}^{\infty} \frac{2^i}{(2^i - \mathbf{E} S_d/u)^2} < \frac{4 \, \text{var} S_n}{u}$$

The proof of the second statement is similar:

$$\mathbf{Pr}(Y \in U) = \mathbf{E}\,\mathbf{Pr}(Y \in U | X_v, S_d) = \mathbf{E} \frac{\sum_{v \in U} X_v}{S_d} \min\left\{\frac{S_d}{u}, 1\right\} \tag{7}$$

$$= \mathbf{E} \frac{\sum_{v \in U} X_v}{S_d} \left(\frac{S_d}{u} - \frac{(S_d - u)\mathbb{1}_{S_d > u}}{u}\right) \tag{8}$$

$$\geq \frac{\sum_{v \in U} \mathbf{E} X_v}{u} - \mathbf{E} \frac{(S_d - u)\mathbb{1}_{S_d > u}}{u} \geq \frac{\sum_{v \in U} \mathbf{E} X_v}{u} - \frac{4 \, \text{var}(S_d)}{u^2} \tag{9}$$

An application of Prop. 3) finishes the proof. The upper bound is obtained by dropping the last term of (8) and applying Prop. 3. □

The *variation distance* between two distributions μ and π on a finite set X is defined as

$$\|\mu - \pi\|_{var} = \max_{A \subseteq X} |\mu(A) - \pi(A)| \tag{10}$$

The generator receives as input the θ_i ($0 \leq i < d$) parameters of the 'Go with the winners' algorithm. We call these parameters *good* if for all i: $\prod_{j=1}^{i} \theta_j \leq ea(i)$. One way to obtain a good set of parameters θ_i is to sample them recursively using algorithm *Sample_θ* as analyzed in [AV94].

Theorem 6. *Let μ be the output distribution of Gwtw-Gen for the following parameters: the desired depth is d, $\theta_1, \ldots, \theta_{d-1}$ are good, $B \geq 2ed\kappa$, and $u \geq 4\,\mathbf{E} S_d$. Then*

$$\|\mu - \pi\|_{var} \leq \frac{4ed\kappa}{B(u/\mathbf{E} S_d)}$$

The expected time needed by Gwtw-Gen to produce an output is $\Theta(Bd(u/\mathbf{E} S_d))$.

Proof. (sketch) Let Z be the output of Gwtw-Gen. For any $U \subseteq V_d$: $\mu(U) = \mathbf{Pr}(Z \in U) = \mathbf{Pr}(Y \in U | Y \neq \perp) = \mathbf{Pr}(Y \in U)/\mathbf{Pr}(Y \neq \perp)$. Thus, the bounds of Lemma 5 yield:

$$\pi(U) - 4\,\text{var}(S_d)/(u\,\mathbf{E} S_d) \leq \mathbf{Pr}(Z \in U) \leq \pi(U)(1 - 4\,\text{var}(S_d)/(u\,\mathbf{E} S_d))^{-1}$$

By [AV94, Lem. 2], $\text{var}(S_d)/(\mathbf{E} S_d)^2 \leq ed\kappa/B$. The theorem follows since $\pi(U) - \mu(U) = \mu(V_d \setminus U) - \pi(V_d \setminus U)$. □

Proposition 7. *For every generation problem $R \subseteq \Sigma^* \times \Sigma^*$ and weight function $wt : \Sigma \mapsto \mathbb{R}$, there exists a 'go with the winners'-based generator which, on input x and $\epsilon > 0$, generates elements of R_x from a distribution μ_x such that*

$$\|\mu_x - \pi_x\|_{var} < \epsilon$$

where π_x is the wt-multiplicative distribution on R_x. The expected time per generated element is $O(l_x^2 \kappa \epsilon^{-1} t_P)$, where t_P is the time needed to evaluate predicate P.

The proposition follows from Prop. 2 and Thm. 6. Since R is decidable in polynomial time, one can always find P such that t_P is polynomial. Note that P determines the tree and, thus, κ. The remaining task, when applying 'Gwtw-Gen' to a concrete application is to bound κ.

3 Applications

3.1 Self-Avoiding Walks

Under a plausible combinatorial hypothesis, the algorithm of [RS94], generates self-avoiding walks almost uniformly in polynomial time by simulating a Markov chain on a tree. The hypothesis states that for any $i, j \in \mathbb{N}$, the result of concatenating two random self-avoiding walks of lengths i and j is again a self-avoiding walk (of length $i + j$) with a probability lower bounded by a polynomial in $i + j$. Given Theorem 6, a similar result can be shown for Gwtw-Gen and the tree defined in Example 1.

Proposition 8. *There is a 'go with the winners' based generator which produces self-avoiding walks of length d from a distribution μ whose variation distance from the uniform distribution is bounded by $\epsilon > 0$. Given the assumption of [RS94], the running time is bounded by a polynomial in d and ϵ^{-1}.*

The proof is omitted due to space limitations. The time bound compares unfavorably with that of [RS94] w.r.t. ϵ. The algorithm of [RS94] runs in time polynomial in $\log \epsilon^{-1}$. On the other hand, the condition needed to prove Proposition 8 is strictly weaker than that used in [RS94].

3.2 The Molecular Modeling Problem

We aim for a general method rather than for a solution for any particular set of constraints. Our first step will be to transform the problem such that the particular geometric constraints are no longer explicit in the problem definition. We reduce the molecular modeling problem to a weighted tiling problem or general lattice model [Dob68]. The tiling problem is NP-complete [GJ79, Lev73, Lev86]. By the results of Saxe [Sax79] and Yemini [Yem79], the same is true in the worst case for the molecular modeling problem. For this reason, it will be necessary to restrict the constraints and distributions severely, if we are to obtain a polynomial time algorithm (cf. remark at the end of this section).

Recall the definition of $\mathcal{H}(n)$ from the introduction. Given an input n (the size of the square) the set $\mathcal{H}(n)$ takes the role of R_x, the set of feasible solutions. Note that membership in $\mathcal{H}(n)$ can be decided locally: (G, f) is in $\mathcal{H}(n)$ if any given atom has the required degree, and the edges incident on the atom have the required lengths and angles. This can be decided locally by considering the d_{max}-ball around each atom separately. We require a similar property from the distribution: The distributions of interest can be written in the form

$$\pi_n(h) = \frac{1}{Z} \prod_e g_l(\|e\|) \prod_{e_1, e_2} g_a(\angle(e_1, e_2)), \quad h \in \mathcal{H}(n) \tag{11}$$

where Z is the normalizing factor, the first product goes over all edges and $g_l(\|e\|)$ depends only on the length $\|e\|$ of e. The second product goes over all pairs of edges which share one vertex and $g_a(\angle(e_1, e_2))$ depends only on the angle $\angle(e_1, e_2)$ between e_1 and e_2. The concrete value assigned to any particular edge length by g_l and to any particular angle by g_a depend on the chemical properties of the substance to be modeled and may vary.

We may write π instead of π_n in order to increase readability. The task is to generate elements of $\mathcal{H}(n)$ from distribution π. If the method developed in the previous section is to be used, the problem must be encoded such that π is a wt-multiplicative distribution for appropriate Σ and wt.

Reduction to Weighted Tiling: A *tile* x is a unit square each of whose four sides is labeled with a color and whose center is labeled by a number (weight) $W(x)$. A tiling of the $n \times n$ square is *legal* if any two neighboring tiles have the same color at the side which forms the border between them. Given a set T of tiles, let $\mathcal{T}_n$ denote the set of legal tilings of the $n \times n$ square. We consider the following probability distribution on $\mathcal{T}_n$:

$$w(t) = \frac{\prod_{i=1}^{n^2} W(t_i)}{\sum_{s \in \mathcal{T}_n} \prod_{i=1}^{n^2} W(s_i)} \tag{12}$$

where $t \in \mathcal{T}_n$ and t_i, s_i is the i-th tile of tiling t, s, respectively. The weighted tiling problem is given as follows: The input is a set T of tiles and a natural number n. The task is to generate legal tilings $t \in \mathcal{T}_n$ under distribution $w(t)$.

Given any ordering of the tile positions and choosing $\Sigma = T$, any tiling is a string of n^2 alphabet symbols. Furthermore, w is the W-multiplicative distribution on $\mathcal{T}_n$. Thus, the 'go with the winners' generator can be applied to this problem. We defer the question about the efficiency of the resulting algorithm and turn to the relation between the weighted tiling problem and the original molecular modeling problem.

The following lemma shows that there is a one-to-one relation between tilings (over a certain tile set T) and elements of $\mathcal{H}(n)$ which preserves the distribution. The lemma can be generalized. In fact, the only property of $\mathcal{H}(n)$ it depends on is local decidability.

Lemma 9. *There exists a set of tiles T of constant size such that for each $n \in \mathbb{N}$ there exists a bijection b between $\mathcal{H}(2d_{\max}(n+1))$ and $\mathcal{T}_n$. Furthermore, for every distribution π given by (11), the tile weights $(W(x))_{x \in T}$ can be chosen such that for every $n \in \mathbb{N}$, $t \in \mathcal{T}_n$: $w(t) = \pi(b(t))$.*

Proof. We define a set of tiles and show that it has the properties claimed in the lemma. For each $h \in \mathcal{H}(4d_{\max})$, let there be a be a tile t_h in T. The color on the left, right, top, bottom side of t_h is given by the restriction of h to its left, right, top, bottom half, respectively. In other words, the restriction of h to the subgraph whose vertices are contained in the $4d_{\max} \times 2d_{\max}$ rectangle covering the left half of h is used as a label (color) for the left side of the tile t_h.

To see that the number of tiles is constant, recall that the vertex coordinates lie on a constant size grid. Each $h \in \mathcal{H}(4d_{\max})$ contains only constantly many grid points, and there are only constantly many ways to place vertices and edges on them.

We show the existence of a bijection between $\mathcal{H}(2d_{\max}(n+1))$ and $\mathcal{T}_n$ by constructing an injective mapping from $\mathcal{H}(2d_{\max}(n+1))$ to $\mathcal{T}_n$ and another one from $\mathcal{T}_n$ to $\mathcal{H}(2d_{\max}(n+1))$. Consider any structure $h \in \mathcal{H}(2d_{\max}(n+1))$. Cover it with $4d_{\max} \times 4d_{\max}$ squares whose centers are located at positions $(2d_{\max}i, 2d_{\max}j)$, where $0 < i, j < n$. Let each square be filled with the subgraph (of h) induced by the vertices of h it covers. Thus, each covering square corresponds to an element of $\mathcal{H}(4d_{\max})$ and, therefore, to a tile. In other words, the covering squares define an $n \times n$ tiling. To see that the tiling is legal, it is sufficient to note that neighboring squares overlap and, ignoring edges crossing the boundary, coincide on the $2d_{\max} \times 4d_{\max}$ rectangle which defines the color on the adjacent sides of the corresponding tiles. Thus, we have a mapping from $\mathcal{H}(2d_{\max}(n + 1))$ to $\mathcal{T}_n$. It is injective since any $g, h \in \mathcal{H}(2d_{\max}(n + 1))$ such that $g \neq h$, must differ in at least one vertex position or edge. However, every vertex and every edge are completely covered by at least one square. This is obvious for vertices. For edges it follows from the fact that no edge is longer than $d_{\max}$ and that, due to the overlapping of the covering squares, for every point (except near the boundary of the square containing h), there is a covering square completely containing the $d_{\max}$-ball around that point. The inverse mapping is based on the same principle. Its description is omitted due to space restrictions.

It remains to define the tile weights W. Let t be a tile and let $h_t \in \mathcal{H}(4d_{\max})$ be the structure it corresponds to. Let d_i denote the i-th factor $g_l(\|e\|)$ or $g_a(\angle(e_1, e_2))$ in the numerator of (11). The basic idea is to assign to t the product of the d_i which correspond to the contributions made by vertices and edges of h_t. A slight modification is necessary as, due to the overlapping of the tiles, some contributions are assigned to several tiles. Contributions d_i made by vertices and edges contained in a single quadrant of h_t appear in four tiles (the four tiles which overlap on that quadrant). Similarly, contributions d_i whose vertices are contained in the left (right, upper, lower) half, but not in a single quadrant, contribute to two tiles. All other contributions appear in only one tile. We compensate for these multiple appearances of certain contributions by reducing their impact on the individual tiles correspondingly: Set $W(t) = \prod_i d_i'$, where i runs over all contributions made by h_t, and $d'_i = \sqrt[4]{d_i}$ for contributions contained in a single quadrant of h_t; $d'_i = \sqrt{d_i}$ for contributions contained in the left (right, upper, lower) half of h_t; $d'_i = d_i$ for all other contributions. The claimed relation between π and w follows readily. The fact that these numbers need not be rational will introduce a very small bias in the output distribution of the final algorithm. □

One could try to solve the problem directly by applying Gwtw-Gen to the tree construction of Prop. 2, and converting the the outputs of Gwtw-gen (tilings) into elements of $\mathcal{H}(n)$ by means of the construction in the proof of Lem. 9. It appears unlikely that this algorithm would be efficient as that would require a tree with polynomially bounded κ. As mentioned before, the molecular modeling problem as well as the tiling problem are NP-hard. We have to restrict the problems severely, in order to locate the restricted versions 'safely' in P. We have analyzed restricted versions of the problem as well as refinements of the basic algorithm. The result is a polynomial time algorithm for appropriately restricted versions of the problem. Due to space restrictions this part of the result has to be deferred to the final version.

4 Conclusions and Open Problems

We have developed the general framework for a new type of generation algorithm which is based on the 'go with the winners' algorithm. As the generator can, in principle, be applied to a wide range of problems, it is interesting to compare it with the only other known general-purpose generation method: the simulation of Markov chains. At first sight, the 'go with the winners' generator compares poorly with rapidly mixing Markov chains. While for rapidly mixing Markov chains the running time is normally poly-logarithmic in the accuracy ϵ, it is polynomial in ϵ^{-1} for the 'go with the winners' generator. This appears to be a consequence of the fact that the only information available to bound the running time is a second moment estimate (κ).

On the other hand, bounding a second moment estimate like κ can be easier than showing rapid mixing for a Markov chain. The self-avoiding walks application is a concrete example of this. The hypothesis needed to make the 'go with the winners' generator polynomial is strictly weaker that the hypothesis needed to make a structurally similar Markov chain rapidly mixing. It is an interesting problem to search for natural applications for which a 'go with the winners' generator is polynomial, but for which no rapidly mixing Markov chain is known.

Markov chains have the advantage that their transition graph is not a priori restricted to being a tree. 'Go with the winners' generators have the best chance of competing if a tree structure is somehow 'suggested' by the problem definition. In the case of self-avoiding walks certain Markov chains on a tree structure [RS94, BS85] are considered to be among the most efficient ones as it is harder to define and analyze more general transition graphs.

5 Acknowledgments

The first author would like to thank Alistair Sinclair and David Aldous for very helpful discussions.

References

[AV94]   D. Aldous and U. Vazirani. "Go with the winners" algorithms. In *Proceedings 35th IEEE Symposium on the Foundations of Computer Science*, 1994.

[BS85]   A. Berretti and A. Sokal. New monte carlo method for the self avoiding walk. *Journal of Statistical Physics*, 40:483–531, 1985.

[Dob68] P. L. Dobrushin. The description of a random field by means of conditional probabilities and conditions of its regularity. *Theory of Probability and its Applications*, 8(2):197–224, 1968.

[Ell90] S. Elliot. *Physics of Amorphous Materials*. Longman, 1990.

[GJ79]   M. R. Garey and D. S. Johnson. *Computers and Intractability: a guide to the theory of NP-completeness*. W. H. Freeman, 1979.

[GJ97]   L. Goldberg and M. Jerrum. Randomly sampling molecules. In *Proc. of the 8th Annual ACM-SIAM Symposium on Discrete Algorithms*, 1997.

[Gra96] P. Grassberger. The pruned-enriched rosenbluth method: Simulations of theta-polymers of chain length up to 1,000,000. Manuscript, 1996.

[Hen88] B. Hendrickson. Conditions for unique graph embeddings. Technical Report 88-950, Cornell University, Department of Computer Science, 1988.

[JVV86] M. Jerrum, L. Valiant, and V. Vazirani. Random generation of combinatorial structures from a uniform distribution. *Theoretical Computer Science*, 43:169–188, 1986.

[Lev73] L. A. Levin. Universal sorting problems. *Problemy Peredachi Informatsii*, 9(3):265–266, 1973. In Russian.

[Lev86] L. A. Levin. Average case complete problems. *SIAM Journal on Computing*, 15:285–286, 1986.

[RS94]   Dana Randall and Alistair Sinclair. Testable algorithms for self-avoiding walks. In *Proceedings of the Fifth Annual ACM-SIAM Symposium on Discrete Algorithms*, pages 593–602, Philadelphia, Pennsylvania, 23–25 January 1994.

[Sax79] J. Saxe. Two papers on graph embedding problems. Technical Report CMU-CS-80-102, Carnegie-Mellon University, Department of Computer Science, 1979.

[Shi84] A. N. Shiryayev. *Probability*. Springer-Verlag, 1984.

[Sin93] A. Sinclair. *Algorithms For Random Generation And Counting*. Progress In Theoretical Computer Science. Birkhauser, Boston, 1993.

[SJ89]   A. Sinclair and M. Jerrum. Approximate counting, uniform generation and rapidly mixing Markov chains. *Information and Computation*, 82:93–133, 1989.

[Wel93] D. J. A. Welsh. *Complexity: Knots, Colourings and Knots*. Cambridge University Press, 1993.

[Yem79] Y. Yemini. Some theoretical aspects of position-location problems. In *Proceedings 20th IEEE Symposium on the Foundations of Computer Science*, 1979.

Probabilistic Approximation of Some NP Optimization Problems by Finite-State Machines

Dawei Hong[1] , Jean-Camille Birget [*2]

[1] Dept. of Math. & Computer Science, Southwest State University, Marshall, MN
56258, USA hong@ssu.southwest.msus.edu
[2] Dept. of Computer Science & Eng., University of Nebraska, Lincoln, NE
68588-0115, USA, birget@cse.unl.edu

Abstract. We introduce a subclass of NP optimization problems which contains, e.g., Bin Covering and Bin Packing. For each problem in this subclass we prove that with probability tending to 1 as the number of input items tends to infinity, the problem is approximable up to any given constant factor $\varepsilon > 0$ by a finite-state machine. More precisely, let Π be a problem in our subclass of NP optimization problems, and let I be an input represented by a sequence of n independent identically distributed random variables with a fixed distribution. Then for any $\varepsilon > 0$ there exists a finite-state machine which does the following: On a random input I the finite-state machine produces a feasible solution whose objective value $M(I)$ satisfies

$$P\left(\frac{|Opt(I) - M(I)|}{\max\{Opt(I), M(I)\}} \geq \varepsilon \right) \leq K \exp(-hn),$$

when n is large enough. Here K and h are positive constants.

Keywords: NP- optimization problems, approximation, probabilistic algorithms, finite-state machines.

1 Introduction

Definition [9, 2]: An **NP optimization problem** over an alphabet Σ is a four-tuple $\Pi = (\mathcal{I}, \mathcal{S}, m, opt)$, where:
1. $\mathcal{I} \subseteq \Sigma^*$, the set of admissible input instances, is assumed to be recognizable in polynomial time;
2. $\mathcal{S}(I) \subseteq \Sigma^*$ is the set of all feasible solutions on input I, for every $I \in \mathcal{I}$. The relation $\{(I, s) : I \in \mathcal{I}, s \in \mathcal{S}(I)\}$ is assumed to be decidable in deterministic polynomial time.
3. $m: \mathcal{I} \times \Sigma^* \mapsto \mathbf{R}$, the objective function, is a polynomial-time computable function.

[*] Second author's research supported in part by NSF grant DMS-9203981

4. *opt* $\in$ {max, min} indicates whether Π is a maximization or a minimization problem.

We let the inputs be finite sequences of positive rational numbers. We also assume that the values of the objective function m are positive rational numbers. Eventually we will encode rational numbers as strings.

For an NP optimization problem Π we let $Opt(I)$ denote the optimum value of the objective function on input I. Let A be an algorithm which produces a feasible solution with objective value $A(I)$ on input I. We say that Π is approximated by A up to a factor ε iff for any input I we have (see [11])

$$|Opt(I) - A(I)| / \max\{Opt(I), A(I)\} \le \varepsilon.$$

We say that Π is *asymptotically approximated* by A up to a factor ε iff the above relation holds for all inputs $I = (a_1, ..., a_n)$ with n large enough.

We consider approximation properties of certain NP optimization problems in a *probabilistic* setting. We describe the inputs by sequences $(x_1, ..., x_n)$ of independent, identically distributed ("i.i.d.") random variables x_i with values over the positive rationals; the common domain of the x_i is a probability space $(\Omega, \mathcal{B}, P)$ (with underlying set Ω, σ-algebra $\mathcal{B}$, and probability measure $P : \mathcal{B} \to [0,1]$). Note that since each x_i has only positive values, the expectation $E[x_i]$ exists (allowing $+\infty$) and is not zero.

It has often been observed that the probabilistic behavior of an algorithm can be much better than the worst case behavior. Our results illustrate this again. For example, for the Bin Covering problem no approximation algorithm is known with arbitrarily small asymptotic approximation factor, in the worst case (see e.g., [3] and [4] for background). However, we give an algorithm that has these properties with probability tending to 1 (exponentially fast) as the number of input items tends to ∞. In addition, our algorithm is just a finite-state machine (hence, it runs in real time). The fact that here NP-hard problems are approximated by finite-state machines is significant in its own right; however, probability seems to be the key to reasonably good approximations by finite-state machines.

In Section 2 we define our subclass of NP optimization problems. The definition is set up to make $Opt(x_1, ..., x_n)$ have "probabilistic concentration". Nevertheless, the class of NP optimization problems we end up with has a rather natural and simple definition, and contains some well known NP-hard optimization problems (e.g., Bin Packing, Bin Covering). The main part of the paper is the probabilistic analysis, given in Section 3. Our main result is the following:

Theorem. *For any problem Π in our subclass of NP optimization problems and for any $\varepsilon > 0$, there exists a finite-state machine which on a random input $I = (x_1, ..., x_n)$ (consisting of any number $n \ge 1$ of i.i.d. random variables that are positive-rational valued), produces a feasible solution whose objective value $M(I)$ satisfies*

$$P\left(\frac{|Opt(I) - M(I)|}{\max\{Opt(I), M(I)\}} \ge \varepsilon\right) \le K \exp(-hn)$$

when n is large enough. Here $K > 0$ is a universal constant, and $h > 0$ is a constant depending on Π, ε and on the probability distribution of the inputs.

2 A subclass of NP optimization problems

For two sequences I_1 and I_2 with the same number of coordinates, we say that I_1 is *dominated* by I_2 (denoted by $I_1 \preceq I_2$) iff each coordinate in I_1 is less than or equal to the corresponding coordinate in I_2. By $I_1 \cdot I_2$ we denote the *concatenation* of the strings I_1 and I_2. We denote by $\mathbf{Q}$ the set of rational numbers, by $\mathbf{Q}_{>0}$ the set of positive rational numbers, by $\mathbf{S}^*$ the set of all finite sequences of elements of any set $\mathbf{S}$, by $\mathbf{R}$ the set of real numbers, and by $\mathbf{Z}$ the set of integers.

Definition 2.1 *We introduce the following subclass of NP optimization problems. Each problem Π in this class takes finite sequences of positive rational numbers as inputs, either with bounded or unbounded values: $\mathcal{I} = \mathbf{Q}_{>0}^*$ or $\mathcal{I} = \{r \in \mathbf{Q} : 0 < r \leq c\}^*$ (for some rational constant c). We assume the following axioms:*

(A1) *Subadditivity (Superadditivity):*
The empty-string input $I = ()$ is admissible, and $Opt(I) = 0$.

The concatenation of any two admissible input strings I_1 and I_2 is admissible, and the concatenation of two feasible solutions on input I_1 resp. I_2 is a feasible solution on input $I_1 \cdot I_2$.

If Π is a minimization problem then
$Opt(I_1 \cdot I_2) \leq Opt(I_1) + Opt(I_2)$.
If Π is a maximization problem, $Opt(I_1 \cdot I_2) \geq Opt(I_1) + Opt(I_2)$.

(A2) *Monotonicity:*
For any two admissible input strings I_1 and I_2, if $I_1 \preceq I_2$ then $Opt(I_1) \leq Opt(I_2)$.

(A3) *Restriction Axiom:*
If $I = (a_1, ..., a_n)$ is an admissible input with $n > 0$, then $I_{\hat{r}} = (a_1, ..., a_{r-1}, a_{r+1}, ..., a_n)$ (the sequence obtained by dropping a_r) is also admissible.

There is a non-negative constant λ (depending on Π) such that
$Opt(I_{\hat{r}}) \leq Opt(I) \leq Opt(I_{\hat{r}}) + \lambda$.

(A4) *Non-vanishing of Opt:*
For any sequence $(x_n)_{n \geq 0}$ of i.i.d. random variables with admissible values,

$$\liminf_{n \to \infty} \frac{Opt(x_1, ..., x_n)}{n} > 0 \qquad \text{a.s.}$$

(A5) *Permutation invariance:*
For any $n > 0$, any permutation π of $\{1, ..., n\}$, and any admissible input $(a_1, ..., a_n) \in (\mathbf{Q}_{>0})^n$, the permuted input $(a_{\pi(1)}, ..., a_{\pi(n)})$ is admissible, and
$Opt(a_{\pi(1)}, ..., a_{\pi(n)}) = Opt(a_1, ..., a_n)$.

Note that for every $n > 0$, $Opt(x_1, ..., x_n)$ is a random variable if $x_1, ..., x_n$ are random variables (with admissible positive rational values). This is implied by the fact that for all $n > 0$, the restriction $Opt : (\mathbf{Q}_{>0})^n \to \mathbf{Q}$ is a Borel measurable function. (Indeed, every subset of $\mathbf{Q}^n$ is a Borel set because $\mathbf{Q}^n$ is countable and because singletons are closed.)

As a consequence of axiom A3, for every input I with n coordinates,

$$Opt(I) \leq \lambda n \qquad (1)$$

2.1 Examples: Bin Covering, Bin Packing

Bin Covering and Bin Packing are classical NP optimization problems whose decision versions are NP-complete (see e.g., [3], [6], [11]).

Proposition 2.2 *Bin Covering and Bin Packing belong to our subclass of NP optimization problems.*

Proof: Super- (or sub-) additivity (axiom A1), monotonicity (axiom A2), and permutation invariance (axiom A5) are straightforward to check.

Axiom A3 (with $\lambda = 1$): When we remove an input item from a covering then, except for the bin from which this item is removed, all bins are still covered. A similar argument applies to Bin Packing.

Axiom A4 for Bin Covering: For the classical "first-fit" heuristic (see [6], [8]), $\liminf_{n \to \infty} \frac{FirstFit(x_1, ..., x_n)}{n} \geq \frac{1}{2} E[x_1] > 0$ a.s.
A fortiori, this holds for the function Opt.

For Bin Packing, $Opt(x_1, ..., x_n) \geq \sum_{i=1}^{n} x_i$.
By the Strong Law of Large Numbers,
$\lim_n \sum_{i=1}^{n} x_i/n = E[x_1]$ a.s., and we know that $E[x_1] > 0$. $\square$

3 Approximation with high probability

3.1 Probabilistic concentration

Theorem 3.1 *If Π belongs to our subclass of NP optimization problems then there is a strictly positive constant b (depending on Π and on the input distribution) such that*

$$\lim_{n \to \infty} \frac{Opt(x_1, ..., x_n)}{n} = b \qquad a.s.$$

Moreover, for any $\varepsilon_1 > 0$, and for all n large enough,

$$P\left(\left|\frac{Opt(x_1, ..., x_n)}{n} - b\right| \leq \varepsilon_1\right) \geq$$

$$1 - 2\exp\left(-\frac{\varepsilon_1^2 n}{8\lambda^2}\right)$$

where λ is as in axiom A3 of Definition 2.1.

Proof: By applying Theorem 4.2 (in our Appendix) to the super- (or sub-) additive process $\{Opt(x_{s+1}, ..., x_t) : s, t \in \mathbf{Z}, \; s < t\}$ we have
$\lim_{n \to \infty} \frac{1}{n} Opt(x_1, ..., x_n) = b$ a.s.
for some constant b. By axiom A4, b is strictly positive. This proves the limit result.

The above limit result implies
$\lim_{n \to \infty} \frac{1}{n} E[Opt(x_1, ..., x_n)] = b$.
So for all n large enough,

$$\left| \frac{E[Opt(x_1, ..., x_n)]}{n} - b \right| \leq \frac{\epsilon_1}{2}. \tag{2}$$

Now, for any $n > 0$ we construct a *martingale* $X_0, X_1, ..., X_n$ as follows: For $i = 1, ..., n$,

$$X_i = E[Opt(x_1, ..., x_n) \mid x_1, ..., x_i], \quad \text{and}$$
$$X_0 = E[Opt(x_1, ..., x_n)]$$

Hence, by classical properties of the conditional expectation,

$$X_n = Opt(x_1, ..., x_n).$$

For X_{i-1} with $1 \leq i \leq n + 1$ we also have the following explicit formula:
$X_{i-1}(\omega) =$

$$\int_{v_i, ..., v_n \in \Omega} Opt(x_1(\omega), ..., x_{i-1}(\omega), x_i(v_i), ..., x_n(v_n))$$

$$\cdot dF_i(x_i(v_i)) \cdot .... \cdot dF_n(x_n(v_n)).$$

Then by classical properties of integrals,

$$|X_{i-1} - X_i| \leq \int_{v_i, v_{i+1}, ..., v_n \in \Omega}$$

$$|Opt(x_1(\omega), ..., x_{i-1}(\omega), x_i(v_i), x_{i+1}(v_{i+1}), ..., x_n(v_n)) -$$
$$Opt(x_1(\omega), ..., x_{i-1}(\omega), x_i(\omega), x_{i+1}(v_{i+1}), ..., x_n(v_n))|$$

$$\cdot dF_i(x_i(v_i)) \cdot .... \cdot dF_n(x_n(v_n)).$$

Moreover, $Opt(..., a_{i-1}, b_i, b_{i+1}, ...) - Opt(..., a_{i-1}, a_i, b_{i+1}, ...) =$
$Opt(..., a_{i-1}, b_i, b_{i+1}, ...) - Opt(..., a_{i-1}, b_{i+1}, ...) -$
$(Opt(..., a_{i-1}, a_i, b_{i+1}, ...) - Opt(..., a_{i-1}, b_{i+1}, ...))$
$\leq \lambda$,
where the last inequality follows from axiom A3 and the fact that $Opt(...)$ is non-negative.
Hence we have for $i = 1, ..., n$: $|X_{i-1} - X_i| \leq \lambda$.
Letting $t = (\epsilon_1 n)/2$ and $C_i \leq \lambda$ $(1 \leq i \leq n)$ in Azuma's Lemma we obtain

$$P\left(|Opt(x_1, ..., x_n) - E[Opt(x_1, ..., x_n)]| \leq \frac{1}{2}\epsilon_1 n \right) \geq$$

$$1 - 2\exp\left(-\frac{\epsilon_1^2 n}{8\lambda^2} \right).$$

The result now follows by (2). $\square$

3.2 An approximation algorithm

For every problem Π in our subclass of NP optimization problems (Definition 2.1) we present a real-time *deterministic* algorithm that produces an ε-approximation for Π, with high probability. Recall that we represent the inputs by sequences of i.i.d. random variables $(x_1, ..., x_n)$, $n > 0$, defined on a probability space $(\Omega, \mathcal{B}, P)$, with values in $\mathbf{Q}_{>0}$.

Let us fix an asymptotic approximation factor $\varepsilon > 0$ and a probability distribution $F : \mathbf{Q}_{>0} \to [0, 1]$ for x_i (common to all x_i).

We now start the description of our approximation algorithm $\mathcal{A}_{F,\varepsilon}$.

Input of $\mathcal{A}_{F,\varepsilon}$: a sequence $I = (a_1, ..., a_n) \in \mathcal{I}$;

Output of $\mathcal{A}_{F,\varepsilon}$: a feasible solution $S \in \mathcal{S}(I)$ of Π, whose objective value $m(I, S) = \mathcal{A}_{F,\varepsilon}(I)$ satisfies the probabilistic inequality in Theorem 3.2 below.

Recall (Theorem 3.1) that
$\lim_{n \to \infty} Opt(x_1, ..., x_n)/n = b$ a.s., hence
$\lim_{n \to \infty} E[Opt(x_1, ..., x_n)]/n = b$.
Thus for any $\delta > 0$ there exists n such that
$\left| \frac{1}{n} E[Opt(x_1, ..., x_n)] - b \right| < \delta$.

Preprocessing (independent of the input I):
Pick N large enough so that

$$\left| \frac{E[Opt(x_1, ..., x_N)]}{N} - b \right| < \delta, \tag{3}$$

where δ is a (small) constant (depending on ε) that will be determined in the proof of Theorem 3.2. We also choose N so that it is a square.

Let q_F be the quantile transformation of the distribution F (see (10) in the Appendix). For all i ($1 \le i \le$ N) we define

$$\xi_i^{(\text{N})} = q_F \left(\frac{i}{\text{N}} \right).$$

It is an important to note that q_F *is positive-rational-valued*, since F is the distribution of positive-rational-valued random variables x_i (recall that "min" is used in the definition of the quantile transformation). Also, if $x_i \le c$ then q_F is also bounded by c. Hence each $\xi_i^{(\text{N})}$ is a positive rational number (except perhaps $\xi_N^{(\text{N})}$ which could be $+\infty$ if x_i is unbounded); and $\xi_i^{(\text{N})} \le c$ if $x_i \le c$.

Let us consider the set $\{\xi_1^{(\text{N})}, ..., \xi_N^{(\text{N})}\}$ and consider all sequences of length N consisting of elements from this set (so there are $\le$ N^{N} such sequences).

For every such sequence Ξ, used as an input of problem Π, we pick an *optimal* feasible solution $\Sigma(\Xi) \in \mathcal{S}(\Xi)$; its optimal objective value is $Opt(\Xi)$. We arrange the results of this preprocessing into a table $\mathcal{T}$ which, for every Ξ gives $\Sigma(\Xi)$ and $Opt(\Xi)$.

(In principle we don't know how to find N algorithmically from ε and F in Preprocessing. The algorithm below is for a fixed ε and F.)

By the definition of $\xi_i^{(n)}$, $\xi_1^{(N)} \leq \ldots \leq \xi_N^{(N)}$. The algorithm $\mathcal{A}_{F,\varepsilon}$ will be based on the following function ξ from $\mathbf{Q}_{>0}$ to $\{\xi_1^{(N)}, \ldots, \xi_{N-1}^{(N)}, \xi_N^{(N)}\}$:

$$\xi(x) = \min\{\xi_i^{(N)} : x \leq \xi_i^{(N)}, \ 1 \leq i \leq N\}.$$

Algorithm $\mathcal{A}_{F,\varepsilon}$

> for $k := 0$ to $\lfloor \frac{n}{N} \rfloor - 1$ do
>
> $\Xi_k := (\xi(x_{kN+1}), \ldots, \xi(x_{kN+i}), \ldots, \xi(x_{kN+N}))$;
> in the table $\mathcal{T}$, look up $\Sigma(\Xi_k)$;
>
> return $S = \Sigma(\Xi_1) \cdot \ldots \cdot \Sigma(\Xi_{\lfloor \frac{n}{N} \rfloor})$ (concatenation).

The *intuition* for this algorithm is simple: We break the random input $(x_1, \ldots, x_n)$ into $\lfloor \frac{n}{N} \rfloor$ successive (non-overlapping) segments $I_k = (x_{kN+1}, \ldots, x_{kN+N})$, for $k = 0, 1, \ldots, \lfloor \frac{n}{N} \rfloor - 1$. We discard the remainder $(x_{\lfloor \frac{n}{N} \rfloor N+1}, \ldots, x_n)$ since it has length $<$ N which is asymptotically negligible. For every input segment I_k we find a segment Ξ_k by applying ξ to I_k. By the Kolmogorov-Smirnov statistics (see Theorem 4.5), we expect I_k and Ξ_k to have closely related *Opt* values, with high probability. The optimal solutions $\Sigma(\Xi_k)$ for the Ξ_k's have been precomputed, independently of the input (what depends on the input are the Ξ's that are actually picked, and their order). By axiom A1 we can concatenate feasible solutions. The next Theorem shows that the resulting total solution is close to an optimum, asymptotically, with high probability.

It is interesting (although this is not needed here) that I_k and $\xi(I_k)$ have related distributions: We define the following discrete probability distribution F_N concentrated on $\{\xi_1^{(N)}, \ldots, \xi_{N-1}^{(N)}, \xi_N^{(N)}\}$:

$$F_N(t) = \begin{cases} F(\xi_1^{(N)}) & \text{if } t \leq \xi_1^{(N)}, \\ F(\xi_i^{(N)}) & \text{if } \xi_{i-1}^{(N)} < t \leq \xi_i^{(N)}, \ 1 < i \leq N. \end{cases}$$

Let $X = (x_1, \ldots, x_N)$ be a sequence of i.i.d. random variables with distribution F. Then $\xi(X) = (\xi(x_1), \ldots, \xi(x_N))$ is a sequence of i.i.d. random variables with distribution F_N.

Conversely, if Ξ is a sequence of N i.i.d. random variables over $\{\xi_1^{(N)}, \ldots \xi_{N-1}^{(N)}, \xi_N^{(N)}\}$ with distribution F_N, then Ξ is almost surely equal to $\xi(X)$ for some sequence X of N i.i.d. random variables with distribution F.

Theorem 3.2 (Correctness of $\mathcal{A}_{F,\varepsilon}$). *Let Π be in our subclass of NP optimization problems, and let $I = (x_1, \ldots, x_n)$ be a random input, consisting of i.i.d. random variables with only positive rational values and with distribution*

F. Then for any $\varepsilon > 0$, the algorithm $\mathcal{A}_{F,\varepsilon}$ produces a feasible solution whose objective value $\mathcal{A}_{F,\varepsilon}(I)$ satisfies

$$P\left(\frac{|Opt(I) - \mathcal{A}_{F,\varepsilon}(I)|}{\max\{Opt(I), \mathcal{A}_{F,\varepsilon}(I)\}} \geq \varepsilon\right) \leq K\exp(-hn)$$

when n is large enough. Here $K > 0$ is a universal constant, and $h > 0$ is a constant depending on Π, ε and F.

Proof: We prove the Theorem for the case where Π is a maximization problem; the proof is similar for minimization problems.

As we saw in the intuitive motivation of $\mathcal{A}_{F,\varepsilon}$, we can assume for simplicity that n is divisible by N (this has no effect asymptotically). Accordingly we write $\frac{n}{N}$ instead of $\lfloor\frac{n}{N}\rfloor$. We apply the Hoeffding inequality (15), with $X_k = Opt(\Xi_k)$ $(1 \leq k \leq n/N)$, where Ξ_k is as in the algorithm, and $\Sigma(\Xi_k)$ is the precomputed optimal solution of problem Π on input Ξ_k. Since different Ξ_k are obtained in the algorithm by applying the function ξ to non-overlapping I_k's, it follows that the $Opt(\Xi_k)$'s are i.i.d. random variables. In the Hoeffding inequality we let X be $Opt(\Xi_1)$. Since $0 < Opt(\Xi_k) \leq \lambda N$ (as a consequence of axiom A3, see (1)), we let $n := n/N$, $t' = N\varepsilon$ and $c = \lambda N$ in the Hoeffding inequality. Note that $\sum_{k=1}^{n/N} Opt(\Xi_k) = \mathcal{A}_{F,\varepsilon}(I)$.

Then (15) yields for any $\varepsilon_2 > 0$,

$$P\left(\frac{n}{N} \cdot (E[Opt(\Xi_1)] - N\varepsilon_2) \leq \mathcal{A}_{F,\varepsilon}(I)\right) \geq$$

$$1 - \exp\left(-\frac{2\varepsilon_2^2 n}{N\lambda^2}\right). \tag{4}$$

Let us now estimate $E[Opt(\Xi_1)]$ and compare it with $E[Opt(I_1)]$, where $I_1 = (x_1, ..., x_N)$. Suppose that by sorting the elements of I_1 in nondecreasing order we have $x_{i_1} \leq ... \leq x_{i_N}$ (order statistics). Then the Kolmogorov-Smirnov statistics (13) implies that for any s $(1 \leq s \leq \sqrt{N} + 1)$,

$P(E_1 \cap E_2) \geq 1 - ke^{-2(s-1)^2}$, where

E_1 is the event

$\{(x_{i_{(s-1)\sqrt{N}+1}}, ..., x_{i_N}) \succeq (\xi_1^{(N)}, ..., \xi_{N-(s-1)\sqrt{N}}^{(N)})\}$,

and E_2 is the event

$\{(\xi_{(s-1)\sqrt{N}+1}^{(N)}, ..., \xi_N^{(N)}) \succeq (x_{i_1}, ..., x_{i_{N-(s-1)\sqrt{N}}})\}$.

Recall that N is a square.

For all s $(1 \leq s \leq \sqrt{N} + 1)$, consider the event

$F_s = \{|Opt(I_1) - Opt(\Xi_1)| \leq 2\lambda s\sqrt{N}\}$.

Claim: $E_1 \cap E_2 \subseteq F_{s-1}$.

Proof of the Claim: For all $\omega \in E_1 \cap E_2$ and all k with $2(s-1)\sqrt{N} < k \leq N$,

$$x_{i_{k-2(s-1)\sqrt{N}}}(\omega) \leq \xi_{k-(s-1)\sqrt{N}}^{(N)} \leq x_{i_k}(\omega). \tag{5}$$

Applying ξ to $x_{i_k}(\omega)$ in (5) yields (by the definition of ξ)

$$x_{i_{k-2(s-1)\sqrt{N}}}(\omega) \leq \xi^{(N)}_{k-(s-1)\sqrt{N}} \leq \xi(x_{i_k}(\omega))$$

for $2(s-1)\sqrt{N} < k \leq N$.

By axiom A5 (on permutations), this implies that the subsequence $I'_1(\omega)$ of $I_1(\omega)$ consisting of $x_{i_{k-2(s-1)\sqrt{N}}}$ (with $2(s-1)\sqrt{N} < k \leq N$), is dominated by the appropriately permuted subsequence $\Xi'_1(\omega)$ of $\Xi_1(\omega)$ consisting of $\xi(x_{i_k}(\omega))$ (with $2(s-1)\sqrt{N} < k \leq N$).

Hence by axioms A2 and A3,
$Opt(I_1(\omega)) - 2(s-1)\sqrt{N}\lambda \leq Opt(I'_1(\omega))$
$\leq Opt(\Xi'_1(\omega)) \leq Opt(\Xi_1(\omega))$.
Hence, for all $\omega \in E_1 \cap E_2$,

$$Opt(I_1(\omega)) - 2(s-1)\sqrt{N}\lambda \leq Opt(\Xi_1(\omega)). \tag{6}$$

Applying ξ to $x_{i_{k-2(s-1)\sqrt{N}}}(\omega)$ in (5), by a similar argument we have for all $\omega \in E_1 \cap E_2$,

$$Opt(\Xi_1(\omega)) - 2(s-1)\sqrt{N}\lambda \leq Opt(I_1(\omega)). \tag{7}$$

The Claim now follows from (6) and (7).

The Claim, together with the Kolmogorov-Smirnov theorem (13), implies that for any $s \geq 1$,

$$P(F_s - F_{s-1}) \leq P(\overline{F_{s-1}}) \leq P(\overline{E_1 \cap E_2}) \leq ke^{-2(s-1)^2}.$$

By a classical fact about conditional expectations,

$$|E[Opt(I_1)] - E[Opt(\Xi_1)]| \leq$$

$$\sum_{s=1}^{\sqrt{N}+1} E[|Opt(I_1) - Opt(\Xi_1)| \mid F_s - F_{s-1}] \cdot P(F_s - F_{s-1}).$$

But by the definition F_s,
$E[|Opt(I_1) - Opt(\Xi_1)| \mid F_s - F_{s-1}] \leq 2\lambda s\sqrt{N}$. Therefore

$$|E[Opt(I_1)] - E[Opt(\Xi_1)]| \leq$$

$$2\lambda\sqrt{N} \sum_{s=1}^{\sqrt{N}+1} sP(F_s - F_{s-1}) \leq$$

$$2k\lambda\sqrt{N} \sum_{s=1}^{\infty} se^{-2(s-1)^2} \leq L\sqrt{N},$$

for some constant L depending on Π only. Combining this and (4) we obtain for any $\varepsilon_2 > 0$,

$$P\left(n\left(\frac{E[Opt(I_1)]}{N} - \frac{L}{\sqrt{N}} - \varepsilon_2\right) \leq A_{F,\varepsilon}(I)\right) \geq$$

$$1 - \exp\left(-\frac{2\varepsilon_2^2 n}{N\lambda^2}\right).$$

By the choice of N in terms of δ this yields

$$P(H_1 \cdot n \leq \mathcal{A}_{F,\varepsilon}(I)) \geq 1 - \exp\left(-\frac{2\varepsilon_2^2 n}{N\lambda^2}\right), \tag{8}$$

$$\text{where} \quad H_1 = b - \delta - L\sqrt{\frac{1}{N}} - \varepsilon_2.$$

By Theorem 3.1 and (8) we have for any $\varepsilon_1, \varepsilon_2 > 0$,

$$P\left(1 - H_2 \leq \frac{\mathcal{A}_{F,\varepsilon}(I)}{Opt(I)} \leq 1\right) \geq$$

$$1 - 2\exp\left(-\frac{\varepsilon_1^2 n}{8\lambda^2}\right) - \exp\left(-\frac{2\varepsilon_2^2 n}{N\lambda^2}\right), \tag{9}$$

$$\text{where} \quad H_2 = \frac{\delta}{b + \varepsilon_1} + \frac{L}{b + \varepsilon_1} \cdot \sqrt{\frac{1}{N}} + \frac{\varepsilon_1 + \varepsilon_2}{b + \varepsilon_1}.$$

For any $\varepsilon > 0$, since $b > 0$, we may take ε_1 and ε_2 small enough so that the third term of H_2 is less than $\varepsilon/3$. Then we take δ small enough so that the first term of H_2 is less than $\varepsilon/3$. Finally, in Preprocessing we take N to be a large enough square so that (3) holds, and so that the second term of H_2 is less than $\varepsilon/3$. Now $H_2 < \varepsilon$, and the Theorem follows from (9). $\quad\square$

3.3 The finite-state machine

Theorem 3.3 *Let Π be a problem in our class of NP optimization problems and let $I = (x_1, ..., x_n)$ be a random input, consisting of i.i.d. random variables with positive rational values only, and with distribution F.*

Then there exists a constant $b > 0$ such that $\lim_{n \to \infty} \frac{1}{n} Opt(I) = b$ a.s.

*For any $\varepsilon > 0$, there exists a **finite-state machine** $\mathcal{M}_{F,\varepsilon}$ which on input I produces a feasible solution whose objective value $\mathcal{M}_{F,\varepsilon}(I)$ satisfies*

$$P\left(\frac{|Opt(I) - \mathcal{M}_{F,\varepsilon}(I)|}{\max\{Opt(I), \mathcal{M}_{F,\varepsilon}(I)\}} \geq \varepsilon\right) \leq K \exp(-hn),$$

when n is large enough. Here $K > 0$ is a universal constant and $h > 0$ is a constant depending on Π, ε and F.

Proof: Since ε, F, and N are fixed, $\mathcal{T}$ is fixed and finite. So the algorithm $\mathcal{A}_{F,\varepsilon}$ looks like a finite-state machine (i.e., a Mealy machine), except that so far, input sequences consisted of arbitrary positive rational numbers. The inputs of an automaton have to be strings over a finite alphabet.

We choose to represent any positive rational number in the form $a + \frac{b}{c}$ with $a, b, c \in \mathbf{N}$, and $b < c$ where a, b, c are written in *reverse binary* (with least significant bit at the left); moreover, the numbers b and c are represented as two "parallel bit streams" (i.e., two strings of equal length, possibly with leading zeros, lined up bit by bit) over the alphabet $\{\binom{0}{0}, \binom{0}{1}, \binom{1}{0}, \binom{1}{1}\}$. For example, the string $0001 + \binom{1}{1}\binom{0}{1}\binom{1}{0}\binom{0}{0}\binom{0}{1}$ represents the number $8 + \frac{5}{19}$. So the alphabet of $\mathcal{M}_{F,\varepsilon}$ is $\{0, 1, +, \binom{0}{0}, \binom{0}{1}, \binom{1}{0}, \binom{1}{1}, \#\}$, where $\#$ serves as a separator between input items.

For an input item $a + \frac{b}{c}$ (represented as a string), $\mathcal{M}_{F,\varepsilon}$ simultaneously compares this item with the N *fixed* numbers $\xi_i^{(N)}$, and thus computes Ξ_k from the kth input subsegment I_k. A finite automaton can compare a variable rational number $a + \frac{b}{c}$ (represented as a string), with a fixed number $\xi_i^{(N)}$ as follows:

The finite automaton first compares the integral part a of $a + \frac{b}{c}$ with the (fixed) integral part of $\xi_i^{(N)}$ (this is a classical construction; three states suffice for comparison in reverse binary). If the integral parts are equal, the fractional part $\frac{b}{c}$ of the input item is compared next with the (fixed) fractional part (let's call it $\frac{\beta_i}{\gamma_i}$) of $\xi_i^{(N)}$. This is done as follows:

The finite automaton can multiply a variable number (represented in reverse binary) by a fixed number (using a construction similar to the classical binary adder). Thus, given $\frac{b}{c}$ the automaton computes $\frac{\gamma_i b}{\beta_i c}$ (all fractions represented as parallel bit streams in reverse binary). At the same time, the automaton compares $\beta_i b$ and $\gamma_i c$ to see if the input item is $\le$ or $> \xi_i^{(N)}$ (using a three-state comparator for parallel bit streams in reverse binary). $\square$

4 Appendix on Probability Theory

A *subadditive process* is a family of real random variables $\{X_{s,t} : s, t \in \mathbf{Z}, \ s < t\}$ such that:

(1) For all $s < t < u$, $X_{s,u} \le X_{s,t} + X_{t,u}$;
(2) The distribution of $X_{s,t}$ is completely determined by $t - s$;
(3) $X_{0,t}$ has finite expectation and there is a constant α such that for all $t \ge 1$, $E[X_{0,t}] \ge \alpha t$.

By reversing inequalities about X or $E[X]$ in (1) and (3) we obtain a *super-additive process* (see [10]).

Proposition 4.1 *([10]) If $\{X_{s,t} : s, t \in \mathbf{Z}, \ s < t\}$ is a subadditive (or superadditive) process then there is a constant $\gamma \ne \infty$ such that*

$$\lim_{t \to \infty} \frac{E[X_{0,t}]}{t} = \gamma.$$

Theorem 4.2 (Kingman *[10]* **).** *Let $\{X_{s,t} : s, t \in \mathbf{Z}, \ s < t\}$ be a subadditive or superadditive process, and let γ be the same as in Proposition 4.1. Then for all s,*

$$\lim_{t \to \infty} \frac{X_{s,t}}{t - s} = \gamma \quad \text{a.s.}$$

Theorem 4.3 (Azuma's Lemma) [1]. *Suppose the sequence* $X_0, X_1, ..., X_n$ *is a martingale. For each i (1 ≤ i ≤ n), let* $C_i = \sup\{|X_i(\omega) - X_{i-1}(\omega)| : \omega \in \Omega\}$ *and let* $C = \sum_{i=1}^{n} C_i^2$. *Then for all* $t > 0$,

$$P(|X_n - X_0| > t) \leq 2\exp\left(-\frac{t^2}{2C}\right).$$

The **quantile transformation** q_F of a probability distribution F over $\mathbf{R}$ (see [5]) is the function $[0, 1] \to \mathbf{R} \cup \{-\infty, +\infty\}$, defined by

$$q_F(z) = \min\{t \in \mathbf{R} : F(t) \geq z\}. \tag{10}$$

This definition uses min (instead of inf) because F is continuous from the left.

Proposition 4.4 *Let* x_i *(1 ≤ i ≤ k) be i.i.d. real-valued random variables with distribution* F. *Then there exist uniformly distributed i.i.d. random variables* u_i *(1 ≤ i ≤ k) over* $[0, 1]$ *such that for all i,* $x_i = q_F(u_i)$ *a.s.*

The Kolmogorov-Smirnov statistics: Let $(u_{(1)}, \ldots, u_{(n)})$ be the order statistics of a sequence $(u_1, \ldots, u_n)$ of i.i.d. random variables that are uniformly distributed over $[0, 1]$. The following is a famous result, called the Kolmogorov-Smirnov statistics:

Theorem 4.5 *There is a constant* $k > 0$ *such that for all* $s > 0$,

$$P\left(\max_{1 \leq i \leq n} \left\{\left|u_{(i)} - \frac{i}{n}\right|\right\} \geq \frac{s}{\sqrt{n}}\right) \leq ke^{-2s^2}. \tag{11}$$

Applications of the Kolmogorov-Smirnov statistics:

For all $n \geq 1$ and for all $1 \leq i \leq n$, we define $\xi_i^{(n)} = q_F\left(\frac{i}{n}\right)$. Consider the two events

$$G_n^{(1)} = \bigcap_{i=1}^{n} \left\{ q_F\left(\max\{0, \frac{i}{n} - \frac{s}{\sqrt{n}}\}\right) \leq q_F(u_{(i)}) \right\},$$

$$G_n^{(2)} = \bigcap_{i=1}^{n} \left\{ q_F(u_{(i)}) \leq q_F\left(\min\{1, \frac{i}{n} + \frac{s}{\sqrt{n}}\}\right) \right\}.$$

Then by (11) it follows that for any $s > 0$,

$$P(G_n^{(1)} \cap G_n^{(2)}) \geq 1 - ke^{-2s^2}, \tag{12}$$

where k is the universal constant from (11).

Now consider the two events

$$H_n^{(1)} = \bigcap_{i=s\sqrt{n}+1}^{n} \{q_F(u_{(i)}) \geq \xi_{i-s\sqrt{n}}^{(n)}\},$$

$$H_n^{(2)} = \bigcap_{i=s\sqrt{n}+1}^{n} \{\xi_i^{(n)} \geq q_F(u_{(i-s\sqrt{n})})\}.$$

Then (12) implies that

$$P(H_n^{(1)} \cap H_n^{(2)}) \geq 1 - ke^{-2s^2}. \tag{13}$$

Theorem 4.6 (Hoeffding inequality) [7]. *Let X_1, ..., X_n be i.i.d. random variables, and assume $0 \leq X_i \leq c$ $(i = 1, ..., n)$ for some constant $c > 0$. Let X be another random variable with the same distribution as X_1, ..., X_n. Then for all $t > 0$,*

$$P\left(\sum_{i=1}^{n} X_i - nE[X] \leq t\right) \geq 1 - \exp\left(-\frac{2t^2}{nc^2}\right). \tag{14}$$

Replacing X by $2E[X] - X$ and X_i by $2E[X_i] - X_i$ in (14) yields

$$P\left(nE[X] - \sum_{i=1}^{n} X_i \leq t\right) \geq 1 - \exp\left(-\frac{2t^2}{nc^2}\right). \tag{15}$$

Acknowledgement: Both authors would like to thank David Klarner for his kind encouragement.

References

1. K. Azuma, Weighted sums of certain dependent variables, *Tohoku Mathematical Journal* 3 (1965) 357-367.
2. P. Crescenzi and A. Panconesi, Completeness in approximation classes, *Information and Computation* 93 (1991) 241-262.
3. E. G. Coffman, Jr., M. R. Garey, and D.S. Johnson, Approximation algorithms for bin packing – an updated survey. In G. Ausiello, M. Lucertini, and P. Serafini, editors, *Algorithm Design for Computer System Design*, CISM Courses and Lectures no. 284, pp. 49-106, Springer-Verlag 1984.
4. E. G. Coffman, Jr. and G.S. Lueker, *Probabilistic Analysis of Packing and Partitioning Algorithms*, John Wiley & Sons, 1991.
5. P. Gaenssler, *Empirical Processes*, Lecture Note – Monograph Series, Vol. 3, Institute of Mathematical Statistics, Hayward, CA, 1983.
6. S. Han, D. Hong and J. Y-T. Leung, Probabilistic analysis of a bin covering algorithm, *Operations Research Letters* Vol. 18 No. 4 (1995).

7. W. Hoeffding, Probability inequalities for sums of bounded random variables, *Journal of the American Statistical Association,* 58 (1965) 13-30.

8. D. Hong and J. Y-T. Leung, Probabilistic analysis of k-dimensional packing algorithms, *Information Processing Letters* 55 (1995) 17-24.

9. D. S. Johnson, Approximation algorithms for combinatorial problems, *Journal of Computer and System Sciences* 9 (1974) 256-278.

10. J. F. C. Kingman, Subadditive processes, *Lecture Notes in Mathematics* 539, pp. 168 –222, Springer-Verlag, 1976.

11. C. H. Papadimitriou, *Computational Complexity,* Addison - Wesley, 1994.

Using Hard Problems to Derandomize Algorithms: An Incomplete Survey

Russell Impagliazzo[1]

Computer Science and Engineering, UC, San Diego, 9500 Gilman Drive, La Jolla, CA
92093-0114 russell@cs.ucsd.edu *

Abstract. Yao showed how to use a sufficiently secure cryptographic
permutation to construct pseudo-random generators to de-randomize arbitrary randomized algorithms. To do this, he used the fact that the
XOR of independent random instances of a somewhat hard Boolean
problem becomes almost completely unpredictable, a "direct product
lemma". In this survey, we try to sketch various connections between
hard problems, direct product results, and de-randomization of algorithms.

1 Introduction

This paper addresses the relationship between three central questions in complexity theory. First, to what extent can a problem be easier to solve for probabilistic algorithms than for deterministic ones? Secondly, what properties should
a pseudo-random generator have so that its outputs are "random enough" for
the purpose of simulating a randomized algorithm? Thirdly, if solving one instance of a problem is computationally difficult, is solving several instances of
the problem proportionately harder?

Yao's seminal paper ([16]) was the first to show that these questions are related. Building on work by Blum and Micali ([4]) he showed how to build, from
any cryptographically secure one-way permutation, a pseudo-random generator
whose outputs are indistinguishable from truly random strings to any reasonably
fast computational method. He then showed that such a psuedo-random generator could be used to deterministically simulate any probabilistic algorithm with
sub-exponential overhead.

Blum, Micali, and Yao thus showed the central connection between "hardness" (the computational difficulty of a problem) and "randomness" (the utility
of the problem as the basis for a pseudo-random generator to de-randomize
computation).

Moreover, Yao's construction introduced the XOR Lemma, the prototypical
example of a "direct product lemma" (a specific way in which several instances
or a problem are more difficult than a single instance.) The XOR Lemma can be

* Research supported by NSF YI Award CCR-92-570979, Sloan Research Fellowship
BR-3311, grant #93025 of the joint US-Czechoslovak Science and Technology Program, and USA-Israel BSF Grant 92-00043

paraphrased as follows: Fix a non-uniform model of computation (with certain closure properties) and a boolean (one bit) function f. Assume that any algorithm in the model of a certain complexity has a significant probability of failure when predicting f on a randomly chosen instance x. Then any algorithm (of a slightly smaller complexity) that tries to guess the XOR of $f(x_1), ..f(x_n)$ for n random instances $x_1, ..x_n$ won't do significantly better than a random coin toss.

Since then, similar ideas are behind a series of results that make it seem more likely that randomness does not speed computation super-polynomially. On the one hand, if $P = NP$, then $P = BPP$ ([15]), so a very strong "easiness" result could prove the collapse. On the other, sufficiently strong "hardness" results could also prove that randomness does not help ([16],[13],[1], [2]). There are also results that show seemingly unlikely consequences of a strong separation of P and BPP ([3]).

2 Background

In this section, we give some background needed for the talk. For simplicity, we present the results for the non-uniform Boolean circuit model of computation. Some analogs of the results mentioned in this paper hold for other models as well.

2.1 Probabilistic Computation

One can view a probabilistic computation in the following way. In addition to the actual input x, the algorithm then makes a series of random choices, which we group together as the "random input" r. If the algorithm produces a boolean output, we can define the *acceptance probability* on x to be the fraction of random inputs on which the algorithm outputs 1 or accepts. The class RP is defined as the class of probabilistic algorithms whose acceptance probabilities are either 0 or larger than $1/2$, and the class BPP as those whose acceptance probabilities are either at most $1/4$ or at least $3/4$. Clearly, $RP \subseteq BPP$.

Any polynomial distinction between acceptance and rejection can be amplified by running the algorithm many times and outputting the majority answer. A probabilistic algorithm that has a sub-polynomial distinction between acceptance and rejection would not be useful. However, it might be that not all useful probabilistic algorithms are in BPP. The paradox is explained if one considers that algorithms might have clear distinctions between acceptance and rejection on some inputs, but not on all inputs, as required to be a BPP algorithm. Thus, randomness might be useful in heuristics even if $P = BPP$.

2.2 Pseudo-random sets and hitting sets

The usual way of showing results about BPP based on a hardness assumption also show randomness is not useful in heuristics under these assumptions. The idea is to construct a specific set of "random inputs" that have the same average acceptance probability as the set of all strings. Formally, a set P of m bit strings is *pseudo-random* for security S if for any circuit C of size S, $|Prob_{r \in_U P}[C(r) = 1] - |Prob_{r \in_U \{0,1\}^m}[C(r) = 1]| < 1/m$. We say a family of sets $P_m, m \in N$ is $T(m)$-*constructive* if one can output a list of the elements of P_m in time $T(m)$ given m. It is clear that from a $T(m)$ constructive pseudo-random set for security m, one can put $BPP \subseteq DTIME(T^{O(1)}(m))$. (Pick m larger than the running time of the algorithm. Try the algorithm on each member of the set.) So a polynomially constructive pseudo-random set would show $BPP = P$. Moreover, the same argument works for probabilistic heuristics.

Most results in using hardness to de-randomize algorithms actually construct such a pseudo-random set. In the early results where the hardness was crypto-graphic, the goal was even more ambitious. The goal was to construct sets where one could uniformly sample from the set in time much less than the security. This was because they wanted a pseudo-random generator that was much easier for le-gitimate users to use than for attackers to break. However, for de-randomization this additional requirement is not necessary.

A *hitting set* is a seemingly weaker version of pseudo-randomness than the above. H is *hitting* for security S if for any circuit C of size at most S, if the probability that $C(r) = 1$ for a random string is at least $1/m$, then there is at least one string $h \in H$ so that $C(h) = 1$. It is clear that a hitting set de-randomizes RP. However, a more surprising result is the following:

Theorem 1. *[1] If there is a polynomial-time constructive hitting set for size m, then there is a polynomial-time constructive pseudo-random set.*

As a first step, the authors prove:

Lemma 2. *If there is a polynomial-time constructive hitting set for size m, then there is a boolean function f in $E = DTIME(2^{O(n)})$ that requires circuit com-plexity $C(f) = 2^{\Omega(n)}$.*

2.3 Distributional hardness of functions

In this sub-section, we give the standard definitions of quantitative difficulty of a function for a certain distribution of inputs. The notion of hardness for a function with a small range, with the extreme case being a Boolean function, is slightly more complicated than that for general functions, since one can guess the correct value with a reasonable probability. However, the Goldreich-Levin Theorem ([6]) gives a way to convert hardness for general functions to hardness for Boolean functions.

Definition 3. Let m and ℓ be positive integers. Let $f : \{0,1\}^m \to \{0,1\}^\ell$. $C(f)$, the worst-case circuit complexity of f, is the minimum number of gates for a circuit C with $C(x) = f(x)$ for every $x \in \{0,1\}^m$.

Let π an arbitrary distribution on $\{0,1\}^m$, and let s be an integer. Define the *success* $SUC_s^\pi(f)$ to be the maximum, over all Boolean circuits C of size at most s of $Pr_\pi[C(x) = f(x)]$. Note that if f can be exactly computed in size s (i.e., $C(f) \leq s$), then $SUC_s^\pi(f) = 1$ for any distribution.

For the (important) case $\ell = 1$ we define the *advantage* $ADV_s^x(f) = 2SUC_s^x(f) - 1$. In both notations we replace the superscript π by A if π is the uniform distribution on a subset $A \subseteq \{0,1\}^m$. When π is uniform on $\{0,1\}^m$ we eliminate the superscript altogether.

The *hard-core predicate* result of Goldreich and Levin [6] allows a hardness result for a function with a linear-size output to be converted to a hardness result for a Boolean function.

Definition 4. Let $f : \{0,1\}^m \to \{0,1\}^\ell$. Define $f^{GL} : \{0,1\}^m \times \{0,1\}^\ell \to \{0,1\}$ by $f^{GL}(x,y) =< f(x), y >$, where $< \cdot, \cdot >$ denotes the inner product in $GF(2)$. For a distribution π on $\{0,1\}^m$ define π^{GL} on $\{0,1\}^m \times \{0,1\}^\ell$ be π on the first coordinate and uniform on the second, independently. The function f^{GL} is called the *Goldreich-Levin predicate* for f.

It is easy to see that $ADV_{s+O(m)}^{\pi'}(f^{GL}) \geq SUC_s^\pi(f)$. (Use the inner product of your guess for f and r as the guess for the inner product bit. If your guess for f is correct, so is this guess. If not, the guessed bit is correct with probability $1/2$.) The Goldreich-Levin hard core predicate theorem is a strong converse of this inequality. The following paraphrases the main result of [6] in our notation:

Theorem 5. [6] There are functions $s' = \epsilon^{O(1)}s$, $\delta = \epsilon^{O(1)}$ so that: for every function f and distribution π, if $SUC_s^\pi(f) \leq \delta$, then $ADV_{s'}^{\pi'}(f^{GL}) \leq \epsilon$.

Note that if $\epsilon = 1 - \gamma$, where $\gamma = o(1)$, then $\delta = (1 - \gamma)^{O(1)} = 1 - O(\gamma)$. So between functions and predicates that are only weakly unpredictable the above relationship gives quite tight bounds.

2.4 Hardness versus Randomness

Nisan and Wigderson showed how to use a problem in exponential-time that is sufficiently difficult for non-uniform circuits to simulate randomized computation deterministically.

Theorem 6. [13] If there is a Boolean function f so that $f \in E = DTIME(2^{O(n)})$ and $ADV_{s(n)}(f_n) = 2^{-\Omega(n)}$ for some $s(n) = 2^{\Omega(n)}$, then $P = BPP$.

Their result is actually more general. For example, sub-exponential deterministic simulation of BPP is possible given a function in E with super-polynomial circuit hardness. They give a construction from a hard function of a pseudo-random generator for polynomial-time computation. The seed for the pseudo-random generator has size basically the same as the input size needed for the function to be difficult for circuits of size equal to the time the randomized algorithm takes. The deterministic simulation then runs through all seeds of the pseudo-random generator, and simulates the randomized algorithm using each pseudo-random outputs. It returns the majority answer. The important points to keep in mind are:

- The function must be hard to predict in the probabilistic sense ; i.e., the advantage must be small, comparable to the inverse of the circuit size.
- The *input length* needed to achieve a given level of hardness determines the efficiency of the deterministic simulation.

Yao's XOR lemma and other previously known direct product results handle the first point, but at the expense of the second.

2.5   Random Self-Reducibility

In the previous sub-section, we have seen that our goal is to convert a worst-case hard function into a probabilistically hard predicate. An important first step is to convert a worst- case hard function into a function that is at least somewhat hard distributionally. Random self-reducibility is a technique to show that a function is as hard distributionally as it is in the worst-case. However, the known random self-reducibility arguments increase the input size somewhat and only guarantee that the distributional success is less than one by an inverse polynomial amount, i.e., that the function is hard on a non-negligible fraction of instances. In section ??, we will give a slight modification of the following theorem with tighter input-size bounds. The bulk of the paper will be spent showing how to "amplify" the small amount of hardness guaranteed by the random-self-reducibility argument to the probabilistic hardness needed in the construction of [13] (Theorem 6) without increasing the instance size.

Definition 7. Let $f : \{0,1\}^n \to \{0,1\}^\ell$, $g : \{0,1\}^m \to \{0,1\}^k$. A *many-one reduction* from f to g is a pair of functions $h : \{0,1\}^n \to \{0,1\}^m$ and $h' : \{0,1\}^k \times \{0,1\}^n \to \{0,1\}^\ell$ so that for every $x \in \{0,1\}^n$, $f(x) = h'(g(h(x)), x)$. If f is a functions on the domain $\{0,1\}^*$, let f_n be f restricted to $\{0,1\}^n$ For two such functions f, g and function $l(n) : Z^+ \to Z^+$, we say f is *polynomial-time many-one reducible* to g with *size bound* $l(n)$ if there is a polynomial-time family of reductions h_n, h'_n from each f_n to $g_{l(n)}$. The reduction is *linear-size* if $l(n) = O(n)$.

The following is trivial:

Proposition 8. If f_n is *polynomial-time many-one reducible to* $g_{l(n)}$ then $C(f_n) \leq C(g_{l(n)}) + p(n)$ for some polynomial p.

Definition 9. An *arity-t randomized truth-table reduction* from f to g is a pair of functions $h : \{0,1\}^n \times \{0,1\}^r \to \{0,1\}^{m \times t}, h(x, R) = h_1(x, R), ..h_t(x, R)$ and $h' : \{0,1\}^{n \times t} \times \{0,1\}^n \times \{0,1\}^r \to \{0,1\}^\ell$ with for any $x \in \{0,1\}^n$,

$$Prob_{R \in \{0,1\}^r}[f(x) = h'(h_1(g(h_1(x, R), ...g(h_t(x, R)), x, R)] \geq 2/3$$

. A *random self reduction* for f is a randomized truth-table reduction from f to f so that for every $x \in \{0,1\}^n$ and every $1 \leq i \leq t$, $h_i(x, R) \in_U \{0,1\}^n$. Let f be a function with domain $\{0,1\}^*$. A *polynomial-time random self-reduction of arity* $t(n)$ for f is a polynomial-time computable family of random self-reductions for f_n of arity $t(n)$.

The following well-known fact states that the distibutional and worst-case hardness of random self-reducibile functions are almost identical.

Proposition 10. *If f has a polynomial-time random self-reduction of arity $t(n)$, and $s(n)$ is a function with $SUC_{s(n)}(f_n) \geq 1 - .1/t(n)$, then $C(f) \leq s(n)n^{O(1)}$.*

Note that if the arity of the reduction is large, this gives only a small but non-negligible amount of distributional hardness.

The following result based on ideas of Feigenbaum and Fortnow, [5] shows that random self-reducibility arguments can in fact work for problems in E.

Theorem 11. *For every Boolean function $f \in E$ there is a logarithmic-length function $g \in E$ so that f is linear-size many-one reducible to g and g is polynomial-time random-self-reducible (with arity $O(n^2)$).*

Corollary 12. *For every Boolean function $f \in E$, there is a boolean function $g \in E$ so that $SUC_{C(f_n)n^{-O(1)}}(g_{O(n)}) \leq 1 - O(1/n^2)$. In particular, if $C(f_n) = 2^{\Omega(n)}$, then $SUC_{2^{\Omega(n)}}(g_n) \leq 1 - O(1/n^2)$.*

What we need now is to amplify the distributional hardness of g from being hard on a polynomially small fraction of inputs to being almost indistinguishable from a random bit, as needed for the Nisan-Wigderson construction. Fortunately, there are several results known along these lines.

2.6 Direct Product Lemmas

In the previous sub-sections, we have seen that the main obstacle in de-randomizing BPP based on a hard problem in E is converting a somewhat distributionally hard problem to a probabilistically hard problem without increasing the input size significantly. In this section, we quote the classical Yao XOR lemma, which accomplishes the amplification of hardness, but at the expense of increasing input size. We show its relation to other direct product lemmas, and give a definition of "de-randomized" direct product lemma. We discuss known pseudo-random direct product lemmas.

Let $g : \{0,1\}^n \to \{0,1\}$ and assume $SUC_s(g) \leq 1 - \delta$. In an information theoretic analog, we could thinking of the value of g on a random input as being predictable with probability $1 - 2\delta$ and a random coin flip othehwise. Then the advantage of guessing the exclusive or of k independent copies of g would be $(1 - 2\delta)^k$, since if it were ever a random coin flip, we would be unable to get any advantage in predicting the XOR.

The XOR lemma is a computational manifestation of this intuition. Define $g^{\oplus(k)} : (\{0,1\}^n)^k \to \{0,1\}$ by $g(x_1, \cdots, x_k) = (g(x_1) \oplus \cdots \oplus g(x_k))$.

Theorem 13. *[16],[11]. There are functions k, s' of n, s, ϵ, δ with $k = O(-\log \epsilon/\delta)$, $s' = s(\epsilon\delta)^{O(1)}$ and so that, for any $g : \{0,1\}^n \to \{0,1\}$ with $SUC_s(g) \leq 1 - \delta$, we have: $ADV_{s'}(g^{\oplus(k)}) \leq \epsilon$.*

Theorem 5 reduces the XOR lemma to the following "direct product" theorem, which requires the circuit to output *all* values of $g(x_i)$, rather than their exclusive-or.

Define $g^{(k)} : (\{0,1\}^n)^k \to \{0,1\}^k$ by $g(x_1, \cdots, x_k) = (g(x_1), \cdots, g(x_k))$. Then

Theorem 14. *There are $k = O(-\log \epsilon/\delta)$, $s' = (\epsilon\delta)^{O(1)}s$ so that for any $g : \{0,1\}^n \to \{0,1\}$ with $SUC_s(g) \leq 1 - \delta$, we have: $SUC_{s'}(g^{(k)}) \leq \epsilon$,*

Note that the Goldreich-Levin predicate for $g^{(k)}$ is basically $g^{\oplus k/2}$, (The number of inputs is not fixed at $k/2$ but a binomial distribution with expectation $k/2$; however, if we restrict r in the Goldreich-Levin predicate to have exactly $k/2$ 1's, the bound on the advantage is almost the same.) Thus, combining Theorem 14 with Theorem 5 yields Theorem 13. We'll use this approach in the rest of the paper: first show that computing f on many inputs is hard and then invoke Theorem 5 to make the hard function Boolean.

In [10], the authors give a new proof of Theorem 14, which generalizes to distributions on x which are not uniform and hence can be sampled from using fewer than nk random bits. This allows them to give some conditions which suffice for such a pseudo-random distribution to have direct product properties similar to the uniform distribution. They give a construction of such a pseudo-random generator which (for certain parameters of interest) uses only a linear number of bits to sample n n-bit inputs. We formalize the notion of a direct product result holding with respect to a pseudo-random generator as follows:

Definition 15. Let G be a function from $\{0,1\}^m$ to $(\{0,1\}^n)^k$. G is an $(s, s', \epsilon, \delta)$ direct product generator, if for every boolean function g with $SUC_s(g) \leq 1 - \delta$, we have:

$$SUC_{s'}(g^k \circ G) \leq \epsilon$$

. (Here, the distribution is over uniformly chosen inputs to G.)

Let $s, s', \epsilon, m, \delta$ and k be fixed functions of n, and let G_n be a polynomial-time computable function from $\{0,1\}^m$ to $(\{0,1\}^n)^k$. The family of functions G is called a $(s, s', \epsilon, \delta)$ *direct product generator* if each G_n is a $(s(n), s'(n), \epsilon(n), \delta(n))$ direct product generator.

In this notation, [9] proved:

Theorem 16. *There is a function G from $\{0,1\}^{2n}$ to $(\{0,1\}^n)^n$ which, for any $s = s(n)$, $c \geq 1$, is a $(s, sn^{-O(1)}, O(n^{-(c-1)}), n^{-c})$ direct product generator.*

Combining this result with Theorem 5 allows us to only consider the case $\delta = \Theta(1)$ in the remaining. That this is without loss of generality is stated formally as:

Corollary 17. *Let f be a boolean function on inputs of length n with $SUC_s(f) \leq 1 - n^{O(1)}$. Then there is a boolean function g on inputs of size $O(n)$ so that $SUC_{sn^{-o(1)}}(g) \leq 2/3$ and g is computable in polynomial-time with an oracle for f.*

Corollary 18. *For any Boolean function $f \in E$, there is a Boolean function $g \in E$ with $SUC_{C(f_n)^n - o(1)}(g_{O(n)}) \leq 2/3$.*

The main technical contribution of [10] is:

Lemma 19. *For any $0 < \gamma < 1$, there is a $0 < \gamma' < 1$ and a $c > 0$ so that there is a polynomial time $G : \{0,1\}^{cn} \rightarrow \{0,1\}^{n \times n}$ which is a $(2^{\gamma n}, 2^{\gamma' n}, 2^{-\gamma' n}, 1/3)$ direct product generator.*

¿From Lemma 19, and the known results, we can conclude:

Theorem 20. *If there is a Boolean $f \in E$ with $C(f_n) = 2^{\Omega(n)}$, then there is a polynomial-time constructive pseudo-random set, and so $P = BPP$.*

Proof. Let $f \in E$ with $C(f_n) \geq 2^{\Omega(n)}$. Then by Corollary 18, there is a Boolean function $g \in E$ with $SUC_{2^{\Omega(n)}}(g_{O(n)}) \leq 2/3$, or equivalently, $SUC_{2^{\gamma n}}(g_n) \leq 2/3$ for some constant $\gamma > 0$. Let G be as in Lemma /refmain. Let $h_{cn} = g^{(n)} \circ G_{cn}$. Since G is polynomial-time, $h \in E$, and Then $SUC_{2^{\gamma' n}}(h_{cn}) \leq 2^{-\gamma' n}$ by definition of a direct product generator. By Theorem 5, $ADV_{2^{\Omega n}}(h^{GL}) \leq 2^{-\Omega n}$. Then by Theorem 6, $P = BPP$.

3 Summary and Open Problems

The original constructions of pseudo-random generators were motivated by cryptography and hence had stronger properties than needed for de-randomization. The existence of such cryptographic pseudo-random generators was characterized as equivalent to that of a one-way function in [8]. This paper built on a whole sequence of previous work. Similarly, [10] characterizes the conditions under which a polynomial-time constructive pseudo-random set exists : precisely if there is a function in E requiring exponential circuit complexity. Again, this is just the last step in a long chain.

In fact, is it the last step? Although the existence of polynomial-time constructive pseudo-random sets is characterized, similar results are not known for time complexities between polynomial and exponential. This is an interesting open problem.

The second question is wheter again we are being over-ambitious. Do we need such pseudo-random sets to prove $P = BPP$? As we mentioned earlier, such sets prove that randomness is helpful even in heuristics, and don't use the strong requirement that BPP algorithms must behave well on all inputs. However, maybe this issue just shows that $P = BPP$ is the wrong translation of the philosophical question, "does randomness speed computation?" If randomness helped heuristics, we might still want to say that randomized machines are more powerful than deterministic ones. What is the right formulation? (It seems a little late in the game to be re-opening this issue, but maybe it needs to be done.)

The last question is whether we can use hardness vs. randomness results to prove unconditional bounds on the power of BPP. For example, currently it is not known whether $BPP = NE$! If one could somehow *certify* a function was hard, one could use it to create a pseudo-random set in non-deterministic sub-exponential time and prove this. This certifying question is similar to the existence of NP-natural proofs [14].

References

1. A. Andreev, A. Clementi and J. Rolim, "Hitting Sets Derandomize BPP", in *XXIII International Colloquium on Algorithms, Logic and Programming (ICALP'96)*, 1996.
2. A. Andreev, A. Clementi, and J. Rolim, "Hitting Properties of Hard Boolean Operators and its Consequences on BPP", manuscript, 1996.
3. L. Babai, L. Fortnow, N. Nisan and A. Wigderson, "BPP has Subexponential Time Simulations unless EXPTIME has Publishable Proofs", *Complexity Theory*, Vol 3, pp. 307–318, 1993.
4. M. Blum and S. Micali. "How to Generate Cryptographically Strong Sequences of Pseudo-Random Bits", *SIAM J. Comput.*, Vol. 13, pages 850–864, 1984.
5. J. Feigenbaum and L. Fortnow, "Random self-reducibility of complete sets", SIAM J. of Computing, vol. 22, no. 5 (Oct. 1993). pp. 995-1005.
6. O. Goldreich and L.A. Levin. "A Hard-Core Predicate for all One-Way Functions", in *ACM Symp. on Theory of Computing*, pp. 25–32, 1989.
7. [GNW] O. Goldreich, N. Nisan and A. Wigderson. "On Yao's XOR-Lemma", available via www at ECCC TR95-050, 1995.
8. J. Hastad, R. Impagliazzo, L.A. Levin and M. Luby, "Construction of Pseudo-random Generator from any One-Way Function", to appear in *SICOMP*. (See preliminary versions by Impagliazzo et. al. in *21st STOC* and Hastad in *22nd STOC*.)
9. R. Impagliazzo, "Hard-core Distributions for Somewhat Hard Problems", in *36th FOCS*, pages 538–545, 1995.
10. R. Impagliazzo and A. Wigderson, "$P = BPP$ Unless E has Sub-exponential circuits: De-randomizing the XOR Lemma", STOC '97, pp. 220-229.
11. L.A. Levin, "One-Way Functions and Pseudorandom Generators", *Combinatorica*, Vol. 7, No. 4, pp. 357–363, 1987.
12. N. Nisan, "Pseudo-random bits for constant depth circuits", *Combinatorica 11* (1), pp. 63-70, 1991.
13. N. Nisan, and A. Wigderson, "Hardness vs Randomness", *J. Comput. System Sci.* 49, 149-167, 1994
14. S. Rudich, "NP-Natural Proofs", this conference, to appear.
15. M. Sipser, "A complexity-theoretic approach to randomness". STOC '83, pp. 330-335.
16. A.C. Yao, "Theory and Application of Trapdoor Functions", in *23st FOCS*, pages 80–91, 1982.

Weak and Strong Recognition by 2-way Randomized Automata

Andris Ambainis[1] and Rūsiņš Freivalds[1] and Marek Karpinski[2]

[1] Institute of Mathematics and Computer Science, University of Latvia, Raina bulv. 29, Riga, Latvia[***]

[2] Department of Computer Science, University of Bonn, 53117, Bonn, Germany[†]

Abstract. Languages weakly recognized by a Monte Carlo 2-way finite automaton with n states are proved to be strongly recognized by a Monte Carlo 2-way finite automaton with $n^{O(n)}$ states. This improves dramatically over the previously known result by M.Karpinski and R.Verbeek [10] which is also nontrivial since these languages can be nonregular [5]. For tally languages the increase in the number of states is proved to be only polynomial, and these languages are regular.

1 Introduction

In most of papers on Computational Complexity the authors do not distinguish between the complexity of the recognition of a language and its complement. This comes in a deep contrast with the Recursion Theory where the distinction between recursive and recursively enumerable languages is one of the corner-stones of the theory.

We consider 2-way finite automata and the number of their states as a complexity measure. This is a nice complexity measure reminding space complexity very much.

We follow the terminology introduced by A. Szepietowski [17] and say that a language L is weakly recognized by an automaton A if every input word in the language L is accepted by the automaton and no other input word is accepted. It is allowed that the automaton does not stop on an input word not in the language.

We say that a language L is strongly recognized by an automaton A if every input word in L is accepted by the automaton and every input word not in L is rejected by the automaton.

It has been proved long ago by M. Rabin [15] and J. Shepherdson [16] that every language recognized by nondeterministic or deterministic 2-way finite automata is regular, i.e. the language is recognized by a deterministic 1-way finite automaton as well. They considered only the strong recognition of languages but

[***] Research supported by Grant No.96.0282 from the Latvian Council of Science

[†] Research partially supported by the International Computer Science Institute, Berkeley, California, by the DFG grant KA 673/4-1, and by the ESPRIT BR Grants 7079 and ECUS030

their arguments are valid for the weak recognition as well. On the other hand, R.Freivalds proved in [5] that there is a nonregular language strongly recognized by a Monte Carlo 2-way finite automaton.

M.Karpinski and R.Verbeek [10] proved that if a language is weakly recognized by a Monte Carlo 2-way finite automaton with n states then the language can be strongly recognized by a Monte Carlo 2-way finite automaton with $2^{2^{n^k}}$ states.

We start by proving a better bound for the same problem. Namely, we prove that if a language is weakly recognized by a Monte Carlo 2-way finite automaton with n states then the language can be strongly recognized by a Monte Carlo 2-way finite automaton with $n^{O(n)}$ states.

In Section 3 we concentrate on tally languages, i.e. on languages in a single-letter alphabet. Rather often difficult (and sometimes even undecidable) problems become easily decidable for tally languages. For instance, the class of languages recognizable by 1-way nondeterministic pushdown automata is rather complicated, namely, this is the class of the context-free languages. On the other hand, the class of the tally languages recognizable by 1-way nondeterministic pushdown automata contains only regular languages [14].

In some other cases tally languages recognizable by certain classes of automata can still be rather complicated. For instance, King [13] has shown that non-regular languages can be accepted already by 2-head 1-way alternating finite automata.

We prove that if a tally language is weakly recognized by a Monte Carlo 2-way finite automaton with n states then the language is regular and it can be strongly recognized by a Monte Carlo 2-way finite automaton with $poly(n)$ states and by a deterministic 2-way finite automaton with $2^{poly(n)}$ states.

This bound improves also over the theorem by J.Kaņeps in [9] where it was proved that if a tally language is strongly recognized by a Monte Carlo 2-way finite automaton with n states then the language is regular, and it is recognizable by a deterministic 2-way finite automaton with $2^{2^{n^k}}$ states.

2 Arbitrary Input Alphabets

Theorem 2.1 *Let M be a Monte-Carlo 2-way probabilistic automaton with n states and let L be the language recognized by M. There exists a Monte-Carlo 2-way probabilistic automaton with $n^{O(n)}$ states recognizing L and halting with probability 1 on any input.*

Proof. The configuration of the automaton M is described by (w, m, q) where w is the input word, m is the current position of the automaton and q is its state. We say that (w, m, q) is a *trap*, if, starting from this configuration, M stops with probability 0. A trap is *leftmost* if, starting from it, M never reaches a configuration (w, m', q') with $m' < m$ (i.e. never moves left from position m). A trap is *rightmost* if, starting from it, M never reaches a configuration (w, m', q') with $m' > m$. A rightmost trap (w, m, q) is *repeated* if there is a computation in which M, after starting from (w, m, q), reaches a leftmost trap and, after that, reaches (w, m, q) once again.

Lemma 2.1 *Let p_0 be the smallest probability of transition from one state to another in M. If a configuration (w, m', q') is reachable from (w, m, q), then the probability of reaching (w, m', q') from (w, m, q) in $n|w|$ steps is at least $p_0^{n|w|}$.*

Proof. There are at most $n|w|$ possible configurations of M on the word w. Hence, if we take a sequence of configurations from (w, m, q) to (w, m', q') and eliminate all repetitions, we obtain a way of reaching (w, m', q') in at most $n|w|$ steps. The probability of each step is at least p_0 and the probability of reaching (w, m', q') at least $p_0^{n|w|}$. ☒

Lemma 2.2 *With probability 1 M either stops or reaches a trap.*

Proof. Denote with r_k the probability that, after k steps, M has neither stopped nor reached a trap.

If M has not stopped and its current configuration is not a trap, then there is a halting configuration $((w, m', q')$ where q' is a halting state) which is reachable from the current configuration. By Lemma 2.1, M reaches (w, m', q') in at most $n|w|$ steps with the probability of at least $p_0^{n|w|}$. Hence,

$$r_{k+n|w|} \le (1 - p_0^{n|w|})r_k,$$

and

$$r_{i \cdot n|w|} \le (1 - p_0^{n|w|})^i.$$

When $i \to \infty$, $(1 - p_0^{n|w|})^i \to 0$. Hence, $r_i \to 0$, i.e. M stops or gets into trap with the probability of 1. ☒

Lemma 2.3 *If M is in a trap, it reaches a rightmost (leftmost) trap with the probability of 1.*

Proof. If M has not reached a rightmost trap, there is a reachable configuration (w, m', q') such that m' is greater than the current position of M. Among all such configurations, choose one with the greatest m'. Evidently, it is a rightmost trap. By Lemma 2.1, the probability of reaching it in $n|w|$ steps is at least $p_0^{n|w|}$.

Similarly to Lemma 2.2 we can show that the probability that M does not reach a rightmost trap in i steps after reaching a trap tends to 0. ☒

Lemma 2.4 *If M is in a trap, it reaches repeated rightmost trap with the probability of 1.*

Proof. For any infinite computation, we can construct a sequence of rightmost traps $(w, m_1, q_1), \ldots, (w, m_k, q_k), \ldots$.

(w, m_1, q_1) is the first trap reached by M. Further, (w, m_{k+1}, q_{k+1}) is defined as the first rightmost trap reached by M after reaching at least one leftmost trap after (w, m_k, q_k).

Lemma 2.3 implies that, after reaching (w, m_k, q_k), M reaches a leftmost trap with the probability of 1 and, after reaching a leftmost trap, M reaches a rightmost trap with the probability of 1. Hence, for all k, (w, m_k, q_k) exists with the probability 1.

The number of different configurations of M on a fixed input w is finite. Hence, with probability 1 there exist i, j such that $(w, m_i, q_i) = (w, m_j, q_j)$. (w, m_i, q_i) is a repeated rightmost trap. $\boxtimes$

Below, we describe a method for determining whether the automaton is in a repeated rightmost trap.

Let q_1 and q_2 be states of M. We consider the language U_{q_1, q_2} consisting of all words w' such that M, starting from the state q_2 with the head on the last letter of w',

- never stops,
- never tries to move left from the first letter of w', and
- reaches the last letter of w' in the state q_2 with probability greater than 0, but never tries to move right from it.

Lemma 2.5 *Let (w, m, q) be a repeated rightmost trap. There exists an w' such that w' is a substring of w ending with the mth symbol of w and $w' \in U_{q', q}$ for some q'.*

Proof. Let (w, m', q') be a leftmost trap reached after (w, m, q) and w' be the substring of w from the m'^{th} to the mth symbol.

Starting from (w, m', q'), M never goes to the left from m' or to the right from m because (w, m', q') is a leftmost trap and (w, m, q) is a rightmost trap. However, in some computation M returns to (w, m, q) because (w, m, q) is a repeated trap. Hence, $w' \in U_{q', q}$. $\boxtimes$

So, the problem of determining whether M is in a repeated rightmost trap is reduced to finding a substring $w' \in U_{q', q}$. Further, we prove two lemmas about $U_{q', q}$ and use them to construct an algorithm for determining whether M is in a repeated rightmost trap.

Lemma 2.6 *U_{q, q_2} does not contain two words w_1, w_2 such that w_1 is a proper prefix of w_2.*

Proof. Let $w_2 \in U_{q_1, q_2}$ and w_1 be a proper prefix of w_2. Then, working on w_2, M reaches the last letter of w_2, i.e. moves right from the last letter of w_1. Hence, $w_1 \notin U_{q_1, q_2}$. $\boxtimes$

Lemma 2.7 U_{q_1,q_2} *is recognized by a one-way deterministic automaton with $n^{O(n)}$ states.*

Proof. We start with transforming M into a nondeterministic automaton M'. To do it, we remove all probabilities from the automaton M.

Similarly to M, the automaton M' never stops, never tries to move left from the first letter of w', and reaches the last letter of w' with probability greater than 0, but never tries to move right from it, if the input word belongs to U_{q_1,q_2}. We apply usual determinization techniques to M' and obtain a one-way deterministic automaton. ⊠

Let M_{q_1,q_2} denote the deterministic automaton recognizing U_{q_1,q_2}. Consider the automaton A_{q_1,q_2} working as follows:
For each q_0 which is a halting state in U_{q_1,q_2} do:

1. Set $q = q_0$, $l = 1$. Start from the position m.
2. Repeat:
   (a) If $l = 1$, then:
       i. Move left and find the letter a to the left from the current position. Return back to the current position.
       ii. If there is no state r such that, after seeing a in the state r, M_{q_1,q_2} moves right and changes the state to q, set $l = 0$.
       iii. Otherwise, let r_0 be the smallest such state (in some fixed ordering). Move left and set $q = r_0$, $l = 1$.
   (b) If $l = 0$, then:
       i. Find the state q' to which M_{q_1,q_2} would move from q.
       ii. If there is no state r' such that $q < r'$ (in the same ordering) and M_{q_1,q_2} moves from r' to q', then set $q = q'$, $l = 0$ and move right.
       iii. Otherwise, let r' be the smallest such state. Set $q = r'$, $l = 1$.
   until q is q_0 or starting state of M_{q_1,q_2}.

Let P be the set of all sequences $q_i, \ldots, q_m$ such that, for all $j \in \{i, \ldots, m-1\}$ from the state q_j, seeing the jth symbol of the word w, A_{q_1,q_2} moves to the state q_{i+1} and $q_m = q_0$. P describes all ways how the automaton A_{q_1,q_2} can arrive in the state q_0 to the mth symbol of w.

The automaton M_{q_1,q_2} tracks all sequences in P one by one. It moves to the left until it reaches the beginning of the word or until it cannot continue the current sequence further to the left. At any moment, it can restore its previous state because the automaton A_{q_1,q_2} is deterministic and each pair of state and symbol determines the next state uniquely. Hence, when the automaton cannot continue going left, it correctly returns back and tries the next possibility to go left.

Lemma 2.8 *Assume that there exists an w' such that w' is a substring of w ending with the mth symbol of w and $w' \in U_{q_1,q_2}$. Then, after A_{q_1,q_2} stops, q is a starting state of A_{q_1,q_2}.*

Proof. If such w' exists, there is a sequence in P which starts with the starting state and ends with q_0. (It is the sequence of states from the accepting computation of A_{q_1,q_2} on the word w'.)

The automaton M_{q_1,q_2} finds this sequence and, having found its first (starting) state, stops. ◻

Lemma 2.9 *Assume that there does not exist an w' such that w' is a substring of w ending with the mth symbol of w and $w' \in U_{q_1,q_2}$. Then, after A_{q_1,q_2} stops, q is q_0 and A_{q_1,q_2} has returned to the position m.*

Proof. The number of states in A_{q_1,q_2} is finite and the length of sequences in P is bounded by the length of w. Hence, the number of possible sequences in P is finite. None of these sequences starts with the starting state of A_{q_1,q_2} because any sequence starting with the starting state and ending with the accepting state q_0 gives an accepting computation of A_{q_1,q_2}. Hence, M_{q_1,q_2} traces all possible sequences, does not find a sequence starting with the starting state of A_{q_1,q_2} and returns back to the state where it started (i.e. the mth-symbol of w and the state q_0). Then it stops.

M_{q_1,q_2} does not find q_0 earlier, because Lemma 2.6 implies that the automaton A_{q_1,q_2} cannot get into an accepting state twice on the same word and, hence, paths in P contain q_0 only as their last state.

Hence, it stops only after traversing all possible sequences in P. ◻

Hence, we can determine whether the current configuration is a repeated rightmost trap by running A_{q_1,q_2} for all possible q_1 and all possible halting states of A_{q_1,q_2}. If the current configuration is a repeated rightmost trap, there exists an w' such that w' is a substring of w ending with the mth symbol of w and $w' \in U_{q_1,q_2}$ (Lemma 2.5). Lemma 2.8 implies that it is found by A_{q_1,q_2}. If the configuration is not repeated rightmost trap, there is no such w' and all automata A_{q_1,q_2} return to the starting position (Lemmas 2.5 and 2.9).

For determining whether M is in a rightmost repeated trap, we need to store one state of A_{q_1,q_2} (from $n^{O(n)}$ states) and some small additional information. It can be done with $n^{O(n)}$ states.

We modify M so that, after each step, it checks whether it is in a repeated rightmost trap. If it is, it outputs that the input word does not belong to the language. Otherwise, it continues the computation. (It can be continued because all automata A_{q_1,q_2} return to the position where they start and, hence, current position of M is not lost during checking for a rightmost repeated trap.) The number of states increases at most $n^{O(n)}$ times.

The modified automaton stops with the probability 1 because M stops or gets into repeated rightmost trap with the probability 1 (Lemmas 2.2, 2.4). The probability of the answer $w \in U$ remains the same for all w. ◻

This improves over the upper bound of $2^{2^{n^k}}$ by Karpinski and Verbeek[10].

3   Single-Letter Input Alphabets

Theorem 3.1 *If a language L in a single letter alphabet is weakly recognized by a 2-way Monte Carlo finite automaton with n states, then L can be strongly recognized by a 2-way Monte Carlo finite automaton with poly(n) states.*

For definitions of "configuration" and other technical notions needed for the proof of 3.1 see the proof of Theorem 2.1. However, we need more notions. We elaborate something like a classification of states in the theory of Markov chains [11, 12].

We say that a set of configurations

$$S = \{(w, m_1, q_1), (w, m_2, q_2), \ldots, (w, m_n, q_n)\}$$

is a *cycle*, if the probability to leave S equals 0. The *space* of a cycle S is the cardinality of the set $M_S = \{m_1, m_2, \ldots, m_n\}$. A cycle S is a *leftmost cycle* if the set M_S contains 1. S is a *rightmost cycle* if the set M_S contains $|w|$. The automaton can have several distinct leftmost and rightmost cycles.

For the sake of uniformity, we consider every set consisting of a single accepting (rejecting) configuration as a cycle as well.

A cycle S can be a *subcycle* of another cycle S'.

Lemma 3.1 *The following two properties of a cycle are equivalent:*

- *the cycle S does not contain proper subcycles.*
- *for all pairs of configurations $((w, m', q'), (w, m'', q''))$ in the cycle S, there is a number t such that the probability to reach (w, m'', q'') from (w, m', q') in t steps is positive.*

Proof. Obvious.   ☒

Definition 3.1 *A cycle S with the properties described in Lemma 3.1 is called* minimal.

Lemma 3.2 *If S' and S'' are two different minimal cycles for the same input word w, then these cycles contain no common configurations.*

Proof. Implied by Lemma 3.1.   ☒

The crucial distinction between Theorem 2.1 and Theorem 3.1 is

Lemma 3.3 *If a 2-way probabilistic finite automaton in a single-letter alphabet enters a cycle S, then:*

a) *either the space of the cycle does not exceed the number of the states of the automaton,*

b) *or M_S contains both 0 and $|w|$,*

c) *or the cycle contains a proper subcycle.*

Proof. Suppose that S is a minimal cycle.

Now suppose that $(m'' - m') > n$ where n is the number of states of the automaton. Let $(w, m^{(1)}, q^{(1)}) \rightarrow (w, m^{(2)}, q^{(2)}) \rightarrow \ldots \rightarrow (w, m^{(v)}, q^{(v)})$ be a sequence of configurations such that $m^{(1)} = m'$, $q^{(1)} = q'$, $m^{(v)} = m''$, $q^{(v)} = q''$ and for arbitrary i there is a positive probability to reach $(w, m^{(i+1)}, q^{(i+1)})$ from $(w, m^{(i)}, q^{(i)})$ in a single step. We assume that v is the least possible value for such a sequence for the pair $((w, m', q'), (w, m'', q''))$.

Now assume for the contradiction that $M_S = \{m_1, m_2, \ldots, m_n\}$,

$$\min(m_1, \ldots, m_n) > 0$$

and

$$max(m_1, \ldots, m_n) < |w|.$$

Since $(m'' - m') > n$ and, for arbitrary i, $|m^{(i+1)} - m^{(i)}| \leq 1$, it follows that $v > n$. Hence, there exist $1 \leq i < j \leq v$ such that $q^{(i)} = q^{(j)}$. Since v is the least possible value, it follows that $m^{(j)} > m^{(i)}$.

Denote $q^{(i)} = q^{(j)}$ by q_+. Since there is a positive probability to reach $(w, m^{(i)}, q_+)$ in a finite number of steps, and the automaton sees the same symbol in every square of the tape, there is a positive probability to reach $(w, m^{(2j-i)}, q_+)$ from $(w, m^{(j)}, q_+)$ in a finite number of steps. Reasoning in the same way, we get that $(w, m^{(3j-2i)}, q_+)$, $(w, m^{(4j-3i)}, q_+)$, $\ldots$ are also reachable with a positive probability. This contradicts the existence of $\max(m_1, \ldots, m_n) < |w|$. Symmetrically, one can disprove the existence of $\min(m_1, \ldots, m_k) > 0$. $\boxtimes$

Lemma 3.4 *Whatever the input word, the probability of a 2-way probabilistic finite automaton entering a minimal cycle equals 1.*

Proof. Obvious.

Proof of Theorem 3.1 By Lemma 3.4, the automaton enters a minimal cycle with probability 1. By Lemma 3.3, the minimal cycle either contains the two ends of the tape, or it contains only a small number of the squares of the tape, namely, the number of squares of the tape does not exceed the number of the states of the automaton. The cycles of the latter type are completely characterized by the internal state of the automaton. Beyond that Lemma 3.2 shows that different cycles have non-intersecting sets of configurations.

Given a 2-way probabilistic finite automaton M, we construct a simulating 2-way probabilistic finite automaton M' (with no more than polynomial increase of the number of states) which recognizes the same language but stops with probability 1. The automaton M' uses a complete list of the internal states characterizing the "short" minimal cycles. M' simulates M, however:

1) When M enters such a state, M' checks whether or not the distance between the head of the automaton and the end of tape exceeds the number of states of M. If *yes*, then M has entered a "short" minimal cycle, and M' rejects or accepts the input word depending on whether the cycle is rejecting or accepting one. If *no*, then M' goes on simulating M;

2) When M reaches the end of the tape, M' checks which end of the tape it is. The computation by M starts at the leftmost end of the tape. When M reaches the rightmost end of the tape, M' temporarily interrupts the simulation, tosses a coin near every simbol of the input word $2k$ times (where k is the number of states of M) and returns to the same end of the tape. If *all tosses produce 1*, then M' ends the simulation and rejects the input word. *Otherwise*, M' goes on with the simulation till M reaches the leftmost end of the tape. Such temporary interruptions are made alternately at the two ends of the tape.

To prove the correctness of the above-constructed simulating automaton, one needs to notice that if the given automaton is in a cycle (with probability to leave the cycle equaling 0) containing the two ends of the tape, then the simulating automaton stops with probability 1 rejecting the input. If the given automaton enters a "short" cycle (i.e., a cycle with property a) in Lemma 3.3), then the simulation automaton discovers this fact.

Finally, if the given automaton has not yet entered a cycle, then the probability to stop via the additional coin-tossing procedure is neglectably small in comparison with the probability of the given automaton to stop. ☒

Theorem 3.2 *If a language L in a single letter alphabet is weakly recognized by a 2-way Monte Carlo finite automaton with n states, then L can be strongly recognized by a 1-way deterministic finite automaton with $2^{poly(n)}$ states.*

Proof. The proof heavily relies on two ideas. First, the essence of J.Shepherdson's proof of the regularity of the languages recognizable by deterministic finite 2-way automata [16] was a precise understanding what information about a fragment of an input word is obtained by the deterministic finite 2-way automaton processing the given input word. It is a complete list of the reactions of the deterministic finite 2-way automaton to situations when the automaton enters the fragment:

1) from the left-hand side in the state q_1 ,
2) from the left-hand side in the state q_2 ,
3) from the left-hand side in the state q_3 ,
    $\cdots$;

k) from the left-hand side in the state q_k ,
k+1) from the right-hand side in the state q_1 ,
k+2) from the right-hand side in the state q_2 ,
    $\cdots$,

2k) from the right -hand side in the state q_k ,

These reactions may be exiting the fragment to the left or to the right, or acceptance, or rejection.

Our automaton is a randomized one. Hence the reactions are characterized by their probabilities. Since there may be a positive probability of an infinitely long work, possible reactions are also various minimal cycles. Lemmas 3.1, 3.2, 3.3, 3.4 give us complete description of the possibilities.

Second, a technical lemma by C.Dwork and L.Stockmeyer [2] has already proved its usefullness for the proofs of lower space bounds [7].

Since we model computations of randomized automata by Markov chains, we first give some definitions and results about Markov chains. Basic facts about Markov chains with a finite state space can be found, for example, in [11]. We consider Markov chains with a finite state space, say $1, 2, \ldots, s$ for some s. A particular Markov chain is completely specified by its matrix $R = \{r_{ij}\}_{i,j=1}^{s}$ of transition probabilities. If the Markov chain is in a state i, then it next moves to a state j with probability r_{ij}. The chains we consider have a designated starting state, say, the state 1, and some set T of trapping states, so $r_{kk} = 1$ for all $k \in T$. For $k \in T$, let $a(k, R)$ denote the probability that the Markov chain R is trapped in the state k when started in the state 1.

We start with a lemma which bounds the effect of small changes in the transition probabilities of a Markov chain. This lemma has been taken from [2].

Let $\beta \geq 1$. Say that two numbers r and r' are β-close if either (i) $r = r' = 0$ or (ii) $r > 0$, $r' > 0$, and $\beta^{-1} \leq r/r' \leq \beta$. Two Markov chains $R = \{r_{ij}\}_{i,j=1}^{s}$ and $R' = \{r'_{ij}\}_{i,j=1}^{s}$ are β-close if r_{ij} and r'_{ij} are β-close for all pairs i, j.

Lemma 3.5 ([2]) *Let R and R' be two s-state Markov chains which are β-close, and let k be a trapping state of both R and R'. Then $a(k, R)$ and $a(k, R')$ are β^z-close where $z = 2s$.*

Form a Markov chain $M(x)$ corresponding to an input word x, with states $(Left, q_1)$, $(Left, q_2)$, ..., $(Left, q_k)$, $(Right, q_1)$, $(Right, q_2)$, ..., $(Right, q_k)$. These states are interpreted as follows. Our input word x is in a single-letter alphabet. We consider an "internal" fragment of x, containing nearly all the symbols in the word but the two extremal ones. The states of the Markov chain are interpreted as the reactions of the automaton to the situations 1), 2), $\cdots$,2k). Cycles, acceptance and the rejection are trapping states.

We consider a $2k$ -dimensional unit cube representing these probabilities. Lemma 3.5 shows that if the probabilities are close enough, then the probabilities to accept the input word are also close enough. However our automaton is a Monte Carlo one (the cut-point is isolated). Hence fragments with similar probabilities (in the sense of 3.5) are Myhill-Nerode equivalent, i.e. they can be interchanged and the being or not being in the language is not affected. This way we get that no more than $2^{poly(n)}$ words can be non-equivalent. ⊠

Theorem 3.3 *If a language in a single-letter alphabet is nondeterministically weakly recognized by a 2-way finite automaton with k states, then the language is also Monte Carlo strongly recognized by a 2-way finite automaton with poly(k) states.*

Sketch of proof. If a nondeterministic 2-way finite automaton can accept an input word, it can accept it in time not exceeding $2kn$, where n is the length of the input word. The simulating Monte Carlo automaton guesses $2kn$ nondeterministic branchings equiprobably. If the simulated nondeterministic automaton accepts, the Monte Carlo automaton accepts as well. If the nondeterministic automaton does not accept, then the simulating automaton rejects with probability $2^{-const*k*n}$ or returns for another simulation of the nondeterministic automaton with probability $1 - 2^{-const*k*n}$. ⊠

References

1. A.Ambainis, *The complexity of probabilistic versus deterministic finite automata.* Lecture Notes in Computer Science, 1178(1996).
2. C.Dwork, L.Stockmeyer, *Finite state verifiers I: The power of interaction.* Journal of ACM, 39, 4 (1992), 800–828.
3. R.Freivalds, *Probabilistic machines can use less running time.* Information Processing'77, IFIP, North-Holland, 1977, 839-842.
4. R.Freivalds, *Fast probabilistic algorithms.* Lecture Notes in Computer Science, 74 (1979), 57-69.
5. R.Freivalds, *Probabilistic two-way machines.* Lecture Notes in Computer Science, 118(1981), 33-45.
6. R.Freivalds, *Projections of languages recognizable by probabilistic and alternating finite multitape automata.* Information Processing Letters, 13(1981), 195-198.
7. R.Freivalds, M.Karpinski, *Lower space bounds for randomized computation.* Lecture Notes in Computer Science, 820 (1994), 580-592.
8. A.G.Greenberg, A.Weiss, *A lower bound for probabilistic algorithms for finite state machines.* Journal of Computer and System Sciences, 33 (1986), 88-105.
9. J.Kaņeps, *Regularity of one-letter languages acceptable by 2-way finite probabilistic automata.* Lecture Notes in Computer Science, 529 (1991), 287-296.
10. M. Karpinski, R. Verbeek, *On the Monte Carlo Space Constructible Functions and Separation Results for Probabilistic Complexity Classes.* Information and Computation, 75(1987), 178-189.
11. J.G.Kemeny, J.L.Snell, *Finite Markov Chains.* Van Nostrand, 1960.
12. J.G.Kemeny, J.L.Snell, and A.W.Knapp, *Denumerable Markov Chains.* Springer-Verlag, Berlin et al., 1976, 416 pages.
13. K.N.King, *Alternating multihead finite automata.* Lecture Notes in Computer Science, 115 (1981), 506-520.
14. H.R.Lewis, and Ch.H.Papadimitriou, *Elements of the Theory of Computation.* Prentice-Hall, 1981, 466 pages.
15. M.O.Rabin, *Two-way finite automata.* Proc.Summer Institute of Symbolic Logic, Cornell, 1957, 366-369.
16. J.C.Shepherdson, *The reduction of two-way automata to one-way automata.* IBM Journal of Research and Development, 3(1959), 198-200.
17. A.Szepietowski, *Turing Machines with Sublogarithmic Space.* Lecture Notes in Computer Science, 843 (1994), 115+1 pages.

Tally Languages Accepted by Monte Carlo Pushdown Automata *

Jānis Kaņeps, Dainis Geidmanis and Rūsiņš Freivalds

Institute of Mathematics and Computer Science, University of Latvia, Raina bulv.
29, Riga, Latvia

Abstract. Rather often difficult (and sometimes even undecidable) problems become easily decidable for tally languages, i.e. for languages in a single-letter alphabet. For instance, the class of languages recognizable by 1-way nondeterministic pushdown automata equals the class of the context-free languages, but the class of the tally languages recognizable by 1-way nondeterministic pushdown automata. contains only regular languages [LP81]. We prove that languages over one-letter alphabet accepted by randomized one-way 1-tape Monte Carlo pushdown automata are regular. However Monte Carlo pushdown automata can be much more concise than deterministic 1-way finite state automata.

INTRODUCTION

In this paper we consider languages over single-letter input alphabet accepted by 1-way Monte Carlo pushdown automata.

There are many reasons why to investigate languages over a single-letter alphabet separately. In some cases problems difficult for arbitrary input alphabets become much easier for a single-letter alphabet. Many examples of such a kind are described in the monograph on NP-completeness by M.R.Garey and D.S.Johnson [GJ79].

The class of languages recognizable by 1-way nondeterministic pushdown automata equals the class of the context-free languages, but the class of the languages in a single-letter alphabet recognizable by 1-way nondeterministic pushdown automata contains only regular languages [LP81]. Regular languages are those recognized by deterministic 1-way finite automata.

This was a reason why the case of single-letter alphabet has been singled out by many authors. The languages in a single-letter alphabet have got a special name "tally languages".

However not always tally languages are easy. King [Ki81] has proved that non-regular languages can be accepted already by 2-head 1-way alternating finite automata (2-head 1afa). D.Geidmanis [Ge88] has proved that the class of languages accepted by k-head 1afa is very complicated and for $k > 2$ emptiness problem is undecidable. Many examples of this kind can be found in [Pap94].

---

* This project was supported by Latvian Science Council Grant No. 96.0282

In our results an important role belongs to the property of the considered randomized automata and machines to have so-called isolated cut-point. Nowadays most of the authors have switched the terminology introducing the term "Monte Carlo" instead.

We say that a Monte Carlo machine $\mathcal{M}$ recognizes language L if there is a positive constant δ such that:

1. for arbitrary $x \in L$, the probability of the event " $\mathcal{M}$ accepts x " exceeds $1/2 + \delta$,
2. for arbitrary $x \notin L$, the probability of the event " $\mathcal{M}$ rejects x " exceeds $1/2 + \delta$.

We prove below in Section 1 that languages over a single-letter alphabet accepted by randomized 1-way Monte Carlo pushdown automata are regular.

However Monte Carlo pushdown automata are more complicated devices rather than deterministic 1-way finite automata. Hence one can expect some advantages of the randomized machines as well.

Advantages of randomized Turing machines over deterministic and nondeterministic machines have been studied extensively starting with [Fr75] where palindromes were proved to be recognizable by Monte Carlo off-line Turing machines in less time compared with deterministic machines of the same type. Later similar results were obtained for space and reversal complexity for various types of machines [Fr85, KF90]. On the other hand, it is universally conjectured that randomness helps not always. However, these conjectures were not supported by proofs since proving lower bounds for randomized machines had turned out to be much harder than proving lower bounds for deterministic and nondeterministic machines.

Since for tally languages the capabilities of 1-way Monte Carlo pushdown automata and deterministic 1-way finite automata are the same, the advantages of the randomized machines can be only in the size. And indeed, we prove in Section 2 that 1-way 1-tape Monte Carlo pushdown automata can have nearly exponential size advantage over deterministic 1-way finite automata for recognition of tally languages.

1 REGULARITY

R. Freivalds has shown that 1-way Monte Carlo pushdown automata are more powerful than deterministic ones. For example, it is proved in [Fr81] that the following language L_k can be recognized by one-way Monte Carlo one-counter automaton with arbitrarily small error probability ε: $L \subseteq \{a_1, b_1, \ldots, a_k, b_k\}^*$ consists of all words w such that for all $i \in \{1, \ldots, k\}$ the number of entries a_i in the word w equals the number of entries b_i in w. For $k \geq 3$ the language L_k can not be accepted by any one-way nondeterministic pushdown automaton.

However, Monte Carlo pushdown automata are less investigated than nondeterministic and alternating ones. We are not able to characterize the class of tally languages accepted by Monte Carlo multitape pushdown or finite automata. We have only one result in this direction, which is stated in the following theorem:

Theorem 1. *If a language $L \subseteq \{0\}^*$ is accepted by a 1-way Monte Carlo pushdown automaton then L is regular.*

The probability of the word w being accepted by a 1-way Monte Carlo pushdown automaton $\mathcal{U}$ we denote by $\chi_\mathcal{U}(w)$. We say that $\mathcal{U}$ accepts a language $L \subseteq \Sigma^*$ with δ-isolated cutpoint λ ($0 \leq \lambda \leq 1$, $d > 0$) iff $(\forall w \in L)\ \chi_\mathcal{U}(w) \geq \lambda + \delta$ and $(\forall w \in \Sigma^* \setminus L)\ \chi_\mathcal{U}(w) \leq \lambda - \delta$. Since we consider single-letter alphabet in this paper, we can identify the input word with the length of this word. Following this approach, we introduce the notion of a Monte Carlo pushdown automaton without input. The difference between the usual 1-way Monte Carlo pushdown automaton and 1-way Monte Carlo pushdown automaton without input is that 1-way Monte Carlo pushdown automaton without input has the counter instead of the input tape. The counter never can be decreased.

Monte Carlo pushdown automaton without input is a system $\mathcal{U} = \langle Q, q_0, F, \Gamma, \gamma_0, \varphi \rangle$ where

- Q is a finite set of states,
- $q_0 \in Q$ is the initial state,
- $F \subseteq Q \times \Gamma$ is the set of accepting pairs,
- Γ is a finite pushdown alphabet,
- $\gamma_0 \in \Gamma$ is the bottom symbol,
- $\varphi : Q \times \Gamma \times Q \times \{0, 1\} \times (\Gamma \cup \{POP, IDLE\}) \to [0, 1]$ is a transition function such that for arbitrary $q \in Q$ and $\gamma \in \Gamma$

$$\sum_{q' \in Q} \sum_{x \in \{0,1\}} \sum_{\gamma' \in \Gamma \cup \{POP, IDLE\}} \delta(q, \gamma, q', x, \gamma') = 1,$$

and for all $q, q' \in Q$, $x \in \{0, 1\}$ and $\gamma \in \Gamma$

$$\varphi(q, \gamma_0, q', x, POP) = 0,$$
$$\varphi(q, \gamma, q', x, \gamma_0) = 0.$$

Configuration of $\mathcal{U}$ at any moment is a triple (q, n, w) where $q \in Q$ is the current state, $n \in N$ is the value of the counter, and $w \in \gamma_0 \Gamma^*$ is the content of pushdown. Initial configuration of $\mathcal{U}$ is $(q_0, 0, \gamma_0)$.

If the current configuration of $\mathcal{U}$ is $(n, q, w\gamma)$ where $\gamma \in \Gamma$ then it enters a new configuration $(n + x, q', w')$ with probability

$$\begin{cases} \varphi(q, \gamma, q', x, \gamma') & \text{if } w' = w\gamma\gamma' \text{ where } \gamma' \in \Gamma; \\ \varphi(q, \gamma, q', x, IDLE) & \text{if } w' = w\gamma; \\ \varphi(q, \gamma, q', x, POP) & \text{if } w' = w; \\ 0 & \text{otherwise.} \end{cases}$$

We say that $\mathcal{U}$ has accepted the number $n \in N$ if it enters a configuration $(q, n, w\gamma)$ where $\gamma \in \Gamma$ and $(q, \gamma) \in F$.

The probability of the number n being accepted by $\mathcal{U}$ we denote by $\chi_\mathcal{U}(n)$. We say that $\mathcal{U}$ accepts a set $L \subseteq N$ with δ-isolated cutpoint λ ($0 \leq \lambda \leq 1$, $\delta > 0$) if $(\forall n \in L)\ \chi_\mathcal{U}(n) \geq \lambda + \delta$ and $(\forall n \in N \setminus L)\ \chi_\mathcal{U}(n) \leq \lambda - \delta$.

Note that there is some disparity in the definitions of traditional 1-way Monte Carlo pushdown automaton and 1-way Monte Carlo pushdown automaton without input. The usual 1-way Monte Carlo pushdown automaton has the end marker # on the input tape which allows to process the content of pushdown after reading the input word. 1-way Monte Carlo pushdown automaton without input has not such a possibility: the number n being accepted is decided immediately in the moment of reaching a configuration (n, q, w) where $q \in Q$ and $w \in \gamma_0 \Gamma^*$. However, the next lemma shows that 1-way Monte Carlo pushdown automata are not more powerful than 1-way Monte Carlo pushdown automata without input. This allows us to consider 1-way Monte Carlo pushdown automata without input instead of usual 1-way Monte Carlo pushdown automata in the proof of Theorem 1.

Lemma 2. *For arbitrary 1-way Monte Carlo pushdown automaton $\mathcal{M}$ with input alphabet $\{0\}$ there is a 1-way Monte Carlo pushdown automaton without input $\mathcal{U}$ such that*

$$(\forall n \in N)\, \chi_{\mathcal{M}}(0^n) = \chi_{\mathcal{U}}(n).$$

Sketch of proof. If $\mathcal{D}$ is a Monte Carlo finite automaton and q', q'' are states and w is an input word, then we denote by $p_{\mathcal{D}}(q', w, q'')$ the probability the automaton $\mathcal{D}$ entering the state q'' after reading w, provided it's initial state is q'.

The performance of 1-way Monte Carlo pushdown automaton $\mathcal{M}$ after reaching the end marker # can be simulated by a Monte Carlo finite automaton $\mathcal{A}$ with the input word $mi(w)$ where w is the content of pushdown and $mi(w)$ is the mirror image of w. Let Q be the set of states of $\mathcal{A}$. We take a Monte Carlo finite automaton $\mathcal{B}$ with a set of states Q' and fixed initial state q_0 having the following property: for arbitrary $q', q'' \in Q$ there is a subset $S(q', q'') \subseteq Q'$ such that for arbitrary input word w

$$p_{\mathcal{A}}(q', mi(w), q'') = \sum_{q \in S(q', q'')} p_{\mathcal{B}}(q_0, w, q).$$

The existence of such a Monte Carlo finite automaton $\mathcal{B}$ follows from the result by M. Nasu and N. Honda [NH68] asserting that the class of word functions $f : \Gamma^* \to [0, 1]$ realizable by a Monte Carlo finite automaton is closed under the transposition.

The main idea in the construction of the 1-way Monte Carlo pushdown automaton without input $\mathcal{U}$ is as follows: $\mathcal{U}$ simultaneously simulates $\mathcal{M}$ on the input word and $\mathcal{B}$ on the current content of the pushdown. The pushdown of $\mathcal{U}$ consists of two tapes. The first tape is used for simulation of $\mathcal{M}$. The second tape is used to write the current state of $\mathcal{B}$. This construction allows $\mathcal{U}$ to "remember" the previous state of $\mathcal{B}$ after popping the pushdown top symbol.□

The proof of Theorem 1 needs one more lemma.

Lemma 3. *Let*

- $\eta_i, i \in N, \ldots$ *be a Markov chain with a finite set of states H;*

- (η_i, ν_i), $i \in N$, be a Markov chain with denumerable set of states $H \times N$ having the chain η_i, $i \in N$, as a component;
- $Z_1, Z_2, Z_3, \ldots$ be a sequence of events such that
  (a) $P\{(\eta_0, \nu_0) = (h_0, 0)\} = 1$ where $h_0 \in H$ is the initial state of the chain $\eta_i, i \in N$;
  (b) $(\forall i \in N) \ P\{\nu_{i+1} \geq \nu_i\} = 1$;
  (c) there is a function $f : H \times H \times N \to [0,1]$ such that transition probability from $(h,n) \in H \times N$ to $(h', n') \in H \times N$ for $n' \geq n$ equals $f(h, h', n'-n)$;
  (d) there is a function $g : H \times (N \setminus \{0\}) \to [0,1]$ such that for any integers $t \geq 1$, $n > n' \geq 0$ and any event A depending only on η_i and ν_i ($i = 0, \ldots, t-1$) it follows from $P(A \cap \{(\eta_t, \nu_t) = (h, n')\}) > 0$ that $P\{Z_n | A \cap \{(\eta_t, \nu_t) = (h, n')\}\} = g(h, n - n')$.

Then there is an integer $d \geq 1$ such that $\lim_{n \to \infty} (P(Z_{n+d}) - P(Z_n)) = 0$.

Proof is omitted due to the lack of enough space. $\square$

Proof of Theorem 1. Let $\mathcal{M}$ be the given 1-way Monte Carlo pushdown automaton and let $\mathcal{M}$ accept a language $L \subseteq \{0\}^*$ with a δ-isolated cutpoint λ ($0 \leq \lambda \leq 1, \delta > 0$). By lemma 2, there is a Monte Carlo pushdown automaton without input $\mathcal{U} = (Q, q_0, F, \Gamma, \gamma_0, \varphi)$ such that $(\forall n \in N) \ \chi_{\mathcal{M}}(0^n) = \chi_{\mathcal{U}}(n)$. Hence, for any $n \in N$

$$(0^n \in L) \Rightarrow (\chi_{\mathcal{U}}(n) \geq \lambda + \delta) \quad \text{and} \quad (0^n \notin L) \Rightarrow (\chi_{\mathcal{U}}(n) \leq \lambda - \delta).$$

Let $C = Q \times N \times (\gamma_0 \Gamma^*)$ be the set of configurations of $\mathcal{U}$.

For two configurations $c', c'' \in C$ we denote by $\mu(c', c'')$ the probability of $\mathcal{U}$ next configuration being c'', provided it's current configuration is c'. Probabilities $\mu(c', c'')$ are completely determined by transition function φ. One step transition from c' to c'' is possible iff $\mu(c', c'') > 0$.

We denote by Ω the set of all sequences $(c_0, c_1, c_2, \ldots)$ where $(\forall i \in N) \ c_i \in C$, $c_0 = (q_0, 0, \gamma_0)$ and $(\forall i \in N) \ \mu(c_i, c_{i+1}) > 0$. We call Ω the possibility space of $\mathcal{U}$.

The behaviour of $\mathcal{U}$ from the initial configuration can be viewed as a stochastic process $\xi_0, \xi_1, \xi_0, \ldots$, where ξ_t ($t = 0, 1, 2, \ldots$) is the configuration of B in the moment t and $\xi_0 = (q_0, 0, \gamma_0)$. This is a Markov chain with denumerable set of states C and fixed initial state $(q_0, 0, \gamma_0)$. Any probabilistic event concerning the work of $\mathcal{U}$ can be expressed in form $(\xi_0, \xi_1, \xi_2, \ldots) \in S$ where $S \subseteq \Omega$. We identify this event with the set S.

For any $\omega = (c_0, c_1, c_2, \ldots) \in \Omega$ and any $k \in N$ we define a finite sequence $pref_k(\omega) = (c_0, c_1, \ldots, c_k)$.

Using conventional methodology [KSK76] we define probabilistic space $\langle \Omega, \mathcal{B}, P \rangle$ where

- $\mathcal{B} \subseteq 2^\Omega$ (2^Ω is the set of all subsets of Ω) is minimal σ-algebra containing all sets in form

$$S(c_0, c_1, \ldots, c_k) = \{w \in \Omega | pref_k(\omega) = (c_0, c_1, \ldots, c_k)\}$$

where $k \in N$, $c_0 = (q_0, 0, \gamma_0)$ and for $1 \leq i \leq k$ it holds $\mu(c_{i-1}, c_i) > 0$, and

- $P : \mathcal{B} \to [0, 1]$ is a probability on $\langle \Omega, \mathcal{B} \rangle$ such that

$$P(S(c_0, c_1, \dots, c_k)) = \mu(c_0, c_1)\mu(c_1, c_2) \dots \mu(c_{k-1}, c_k)$$

(this requirement determines $P(A)$ for all $A \in \mathcal{B}$).

Let for $n \in N$

$$Z_n = \{(c_1, c_2, c_3, \dots) \in \Omega | (\exists i \in N)(\exists (q, \gamma) \in F)(\exists w \in \Gamma^*)\, c_i = (q, n, w\gamma)\}$$

be the event "the number n is accepted by $\mathcal{U}$". Clearly, $Z_n \in \mathcal{B}$ and it's probability is $P(Z_n) = \chi_{\mathcal{U}}(n)$.

For any configuration $c \in C$, where $c = (q, n, w\gamma)$ and $\gamma \in \Gamma$, we define $word(c) = w\gamma$ and $pair(c) = (q, \gamma)$.

We say that $t \in N$ is a bottom-moment of a sequence $(c_0, c_1, c_2, \dots) \in \Omega$ if $(\forall i \geq t)\, |word(c_i)| \geq |word(c_t)|$. $t \in N$ being the bottom-moment means that after the moment t neither of $word(c_t)$ symbols will be popped out of the pushdown. Clearly $t = 0$ is a bottom-moment.

Obviously, an arbitrary sequence $\omega \in \Omega$ has infinitely many bottom-moments. We define increasing sequence of bottom-moments

$$\tau_0(\omega) < \tau_1(\omega) < \tau_2(\omega) < \dots$$

where $\tau_0(\omega) = 0$ and for any $i \in N$ there is no bottom-moment t such that $\tau_i(\omega) < t < \tau_{i+1}(\omega)$.

For any $\omega = (c_0, c_1, c_2, \dots) \in \Omega$ and $i \in N$ we define the i-th bottom-pair

$$\eta_i(\omega) = pair(c_{\tau_i}).$$

It is easy to see that τ_i, $i = 1, 2, 3, \dots$, and η_i, $i = 1, 2, 3, \dots$, are sequences of random variables on $\langle \Omega, \mathcal{B}, P \rangle$. Note that $pref_k(\omega') = pref_k(\omega'')$ and $\tau_i(\omega') \leq k$ does not imply $\tau_i(\omega') = \tau_i(\omega'')$, i. e. τ_i and η_i depend also on "future" (not only on "past"). In spite of that, it can be proved that $\eta_0, \eta_1, \eta_2, \dots$ is a Markov chain with a finite set of states $H = Q \times \Gamma$ and fixed initial state $h_0 = (q_0, \gamma_0)$.

Let ν_i be the value of the counter in the i-th bottom-moment. It is obvious that $(\eta_0, \nu_0), (\eta_1, \nu_1), (\eta_2, \nu_2), \dots$ is a Markov chain with the set of states $H \times N$ and the initial state $(h_0, 0)$.

Clearly, $(\forall i \in N)\, P\{\nu_{i+1} \geq \nu_i\} = 1$. It is easy to see that transition probability from $(h, n) \in h \times N$ to $(h', n') \in h \times N$ where $n' \geq n$ depends only on h, h' and $n' - n$, i. e. it equals $f(h, h', n' - n)$ where $f : H \times H \times N \to [0, 1]$ is a function.

Obviously, for any integers $t \geq 1$, $n > n' \geq 0$ and any event A depending only on η_i and ν_i ($i = 0, \dots, t - 1$) the conditional probability $P\{Z_n | A \cap \{(\eta_t, \nu_t) = (h, n')\}\}$ depends only on h and $n' - n$, i. e.

$$P(A \cap \{(\eta_t, \nu_t) = (h, n')\}) > 0 \Rightarrow P\{Z_n | A \cap \{(\eta_t, \nu_t) = (h, n')\}\} = g(h, n - n')$$

where $g : H \times (N \setminus \{0\}) \to [0, 1]$ is a function. Hence, by Lemma 3, there is an integer $d \geq 1$ such that $\lim_{n \to \infty}(P(Z_{n+d}) - P(Z_n)) = 0$. Since the cutpoint λ is isolated, this implies $(P(Z_{n+d}) > \lambda) \Leftrightarrow (P(Z_n) > \lambda)$ for all sufficiently large n. It follows from this that L is regular. $\square$

2 CONCISENESS

We have learned in the preceding Section that tally languages recognized by 1-way Monte Carlo pushdown machines are regular. Hence they can be recognized by 1-way deterministic finite automata as well. We show in this Section that the Monte Carlo machines can be much more concise.

Freivalds[Fr82] considered deterministic 1-way finite automata in a single-letter alphabet and proved the following result:

Lemma 4. *[Fr82] For arbitrary positive integer n there exists a Monte Carlo finite automaton with $O(\frac{\log^2 n}{\log\log n})$ states which recognizes whether or not the input word consists of exactly n symbols.*

Sketch of Proof. The states of the automaton are organized in s cycles. The length of every cycle is a prime number, and the smallest possible prime numbers $p_1 = 2$, $p_2 = 3$, $p_3 = 5$,$p_4 = 7$, $p_5 = 11$, $\ldots$, are used. In the beginning of the work the automaton equiprobably chooses one of the cycles. In every cycle there is exactly one accepting state, namely, the state which corresponds to the remainder of the number n modulo the length of the cycle. Normally for every state in the cycle there is exactly one state for the transition. However once in every cycle there is a special possibility (with an extremely small probability of the transition to a trapping rejecting state). The small probability is chosen in such a way that if the length of the input word is smaller than $N = p_1 * p_2 * p_3 * \ldots * p_s$, then the probability to be trapped in the rejecting state is small, but if the length of the input word exceeds $2 * N$, then the probability to be trapped in the rejecting state is close to 1. The main difficulties of the proof are of number-theoretic nature to justify a possibility of such a small (but not too small) probability. □

Lemma 5. *For arbitrary deterministic 1-way pushdown automaton with k internal states recognizing whether or not the input word consists of exactly n symbols, there is a positive constant c such that, for infinitely many positive integers n, it is true that $k \geq n^c$.*

Sketch of Proof. The deterministic 1-way pushdown automaton cannot return to the already read part of the input tape. Hence the length of the already read part of the input tape is to be remembered. Since the remembering can be done only by the internal states and the pushdown store, no more than k latest squares of pushdown store can be touched.

When the head on the input tape has reach the end-marker, the automaton becomes a deterministic finite-state automaton. □

Lemmas 4, 5 imply

Theorem 6. *There is a sequence of languages L_n in a single-letter alphabet such that:*

1. *For arbitrary positive integer n there exists a Monte Carlo finite automaton with $O(\frac{\log^2 n}{\log\log n})$ states which recognizes L_n,*
2. *For arbitrary deterministic 1-way pushdown automaton with k internal states recognizing L_n there is a positive constant c such that, for infinitely many positive integers n, it is true that $k \geq n^c$.*

References

[Am96]   A.AMBAINIS, *The complexity of probabilistic versus deterministic finite automata.* Lecture Notes in Computer Science, 1178(1996).

[Fr75]   R.FREIVALDS, *Fast computation by probabilistic Turing machines.* Proceedings of Latvian State University, 233 (1975), 201–205 (in Russian).

[Fr79]   R.FREIVALDS, *Fast probabilistic algorithms.* Lecture Notes in Computer Science, 74 (1979), 57-69.

[Fr81]   R.FREIVALDS, *Projections of languages recognizable by probabilistic and alternating finite multitape automata.* Information Processing Letters, v.13, 1981, 195–198.

[Fr81a] R.FREIVALDS, *Capabilities of different models of 1-way probabilistic automata.* Izvestija VUZ, Matematika, No.5 (228), 1981, 26–34 (in Russian).

[Fr81b] R.FREIVALDS, *Probabilistic two-way machines.* Lecture Notes in Computer Science, 118(1981), 33-45.

[Fr82]   R. FREIVALDS, *On the growth of the number of states in result of determinization of probabilistic finite automata,* Avtomatika i Vičislitelnaja Tehnika, 1982, N.3, 39-42 (in Russian)

[Fr85]   R.FREIVALDS, *Space and reversal complexity of probabilistic one-way Turing machines.* Annals of Discrete Mathematics, 24 (1985), 39–50.

[FK94]   R.FREIVALDS, M.KARPINSKI, *Lower space bounds for randomized computation.* Lecture Notes in Computer Science, 820 (1994), 580-592.

[GM79]   N. Z. GABBASOV, T. A. MURTAZINA, *Improving the estimate of Rabin's reduction theorem,* Algorithms and Automata, Kazan University, 1979, 7-10 (in Russian)

[GJ79]   M.R.GAREY, D.S.JOHNSON, *Computers and Intractability: A Guide to the Theory of NP-completeness.* Freeman, San Francisco, 1979.

[Ge88]   D.GEIDMANIS, *On the capabilities of alternating and nondeterministic multitape automata.* Proc. Found. of Comp. Theory, LNCS 278, Springer, Berlin, 1988, 150-154.

[Hr85]   J.HROMKOVIČ, *On the power of alternation in automata theory.* Journ. of Comp.and System Sci., Vol.31, No.1, 1985, 28–39

[HI68]   M.A.HARRISON and O.H.IBARRA, *Multi-tape and multi-head pushdown automata.* Information and Control, 13, 1968, 433–470.

[Ka91]   J.KAŅEPS, *Regularity of one-letter languages acceptable by 2-way finite probabilistic automata.* LNCS 529, Springer, Berlin, 1991, 287–296.

[Ka89]   J.KAŅEPS, *Stochasticity of languages recognized by two-way finite probabilistic automata.* Diskretnaya matematika, 1 (1989), 63–77 (Russian).

[Ka91]   J.KAŅEPS, *Regularity of one-letter languages acceptable by 2-way finite probabilistic automata.* Lecture Notes in Computer Science, 529 (1991), 287-296.

[KF90]   J.KAŅEPS, R.FREIVALDS, *Minimal nontrivial space complexity of probabilistic one-way Turing machines.* LNCS, Springer, 452 1990, 355–361.

[KV87]   M.KARPINSKI, R.VERBEEK, *On the Monte Carlo space constructible functions and separation results for probabilistic complexity classes.* Information and Computation, 75 (1987), 178–189.

[KS60]   J.G.KEMENY, J.L.SNELL, *Finite Markov Chains.* Van Nostrand, 1960.

[KSK76]   J.G.KEMENY, J.L.SNELL, A.W.KNAPP, *Denumerable Markov Chains.* Springer-Verlag, 1976.

[Ki81]   K.N.KING, *Alternating multihead finite automata.* LNCS 115, Springer, Berlin, 1981, 506–520.

[LLS78]   R.E.LADNER, R.J.LIPTON, and L.J.STOCKMEYER, *Alternating pushdown automata.* Conf. Rec. IEEE 19th Ann. Symp. on Found. of Comp. Sci. (1978), 92–106.

[LP81]   H.R.LEWIS, and CH.H.PAPADIMITRIOU, *Elements of the Theory of Computation.* Prentice-Hall, 1981, 466 pages.

[NH68]   M.NASU and N.HONDA, *Fuzzy events realized by finite probabilistic automata.* Information and Control, 12, 1968, 284–303.

[Pap94]   CH.H.PAPADIMITRIOU, *Computational Complexity.* Addison-Wesley, 1994.

[Paz66]   A. PAZ, *Some aspects of probabilistic automata,* Information and Control, 9(1966)

[Ra63]   M. O. RABIN, *Probabilistic automata,* Information and Control, 6(1963), 230-245.

[Sh56]   J.C.SHEPHERDSON, *The reduction of two-way automata to one-way automata.* IBM Journal of Research and Development, 3(1959), 198-200.

[TB73]   B. A. TRAKHTENBROT, Y. M. BARZDIN, *Finite Automata: Behaviour and Synthesis.* North-Holland, 1973.

Resource-Bounded Randomness and Compressibility with Respect to Nonuniform Measures

Steven M. Kautz

Department of Mathematics, Randolph-Macon Woman's College, 2500 Rivermont Avenue, Lynchburg, VA 24503 (skautz@rmwc.edu).

Abstract. Most research on resource-bounded measure and randomness has focused on the uniform probability density, or Lebesgue measure, on $\{0,1\}^\infty$; the study of resource-bounded measure theory with respect to a *non*uniform underlying measure was recently initiated by Breutzmann and Lutz [1]. In this paper we prove a series of fundamental results on the role of nonuniform measures in resource-bounded measure theory. These results provide new tools for analyzing and constructing martingales and, in particular, offer new insight into the compressibility characterization of randomness given recently by Buhrman and Longpré [2]. We give several new characterizations of resource-bounded randomness with respect to an underlying measure μ: the first identifies those martingales whose rate of success is asymptotically *optimal* on the given sequence; the second identifies martingales which induce a *maximal compression* of the sequence; the third is a (nontrivial) extension of the compressibility characterization to the nonuniform case. In addition we prove several technical results of independent interest, including an extension to resource-bounded measure of the classical theorem of Kakutani on the equivalence of product measures; this answers an open question in [1].

1 Introduction

Resource-bounded measure theory, as developed by Lutz in the late 1980's, provides a means for quantitatively analyzing the structure of complexity classes as well as for characterizing random or pseudorandom elements within them. Results in this area published over the last seven years include a growing body of new insights into familiar problems in computational complexity; see [8] for a recently updated survey.

Most work in the areas of resource-bounded measure and randomness has focused on the uniform distribution on $\{0,1\}^\infty$, or Lebesgue measure, in which the probabilities associated with the bits of an infinite sequence are *independent* and *uniform*; i.e., the probability that a given bit is "1" is always one-half. The uniform measure unquestionably provides the most natural starting point for developing a theory of measure and for investigating randomized complexity. Moreover, for example, Breutzmann and Lutz have recently shown that the resource-bounded measure of most complexity classes of interest is to a certain extent invariant with respect to the underlying distribution. Nonetheless there are several reasons for considering nonuniform measures.

One reason, of course, is that for modeling and analyzing computation relative to a physical source of randomness both the assumptions of independence and uniformity are likely to be unrealistic. A number of authors have studied feasible computation relative to random sequences with nonuniform distributions, such as quasi-random and slightly-random sources for BPP [13, 18], and more recently "extractors" and "dispersers" (see the survey [12]).

However, what particularly concerns us here is that results on nonuniform measures are essential to an understanding of the subject of randomness as a whole, and consequently yield techniques which can be fruitfully used to obtain results even in the uniform case. Examples of such methods can easily be found in the somewhat more mature area of constructive measure (i.e., algorithmic randomness in the sense of Martin-Löf); to cite just a few examples, techniques involving nonuniform distributions form the basis of van Lambalgen's new proof of Ville's theorem [17] and of Shen's proof that there are Kolmogorov-Loveland stochastic sequences which are not algorithmically random [15]. Indeed, any truth-table reduction induces a computable measure which is, in general, nonuniform; see [6] for details and applications. Although the study of *resource-bounded* measure in the nonuniform case was only initiated recently in [1], related techniques are already in use by Lutz and Mayordomo in [9].

In this paper we present a number of results on nonuniform distributions in the context of resource-bounded measure and the corresponding notions of resource-bounded randomness, beginning with the observation that *any* nontrivial martingale is implicitly constructed from a nonuniform measure representing its "betting strategy." (Martingales and other fundamental notions in resource-bounded measure are discussed in Section 2.) Thus, even if one is ultimately interested only in the uniform case, at the very least the analysis of and construction of martingales can be facilitated by an understanding of the role of this implicit, nonuniform measure. With this understanding in hand, for example, it is easy to show that within a given class of resource bounds Δ, a martingale succeeds *optimally* on a given sequence (i.e., dominates any other martingale on initial segments of the given sequence) just when the sequence is Δ-random with respect to the martingale's betting strategy (see Section 3.1).

One of our main contributions is to offer new insight into the "compressibility" characterization of resource-bounded randomness. During the 1960's Martin-Löf proposed a definition of infinite random sequences based on a form of constructive measure; at the same time, Kolmogorov, Levin, and others proposed a definition based on incompressibility of initial segments, and it was later shown that the two approaches yield exactly the same class of random sequences, the so-called algorithmically random or 1-random sequences. (See [7] for historical details and references.) Recently Buhrman and Longpré [2] have shown that resource-bounded randomness can also be characterized in terms of a kind of compressibility, providing a beautiful unification of the two approaches to randomness in the resource-bounded setting. Very briefly, they show that if a martingale succeeds on a sequence A, it is possible to define a "compressed" sequence B along with algorithms for transforming A to B and vice versa (the "compression" and "decompression", respectively) which are induced in a uniform way by the martingale. In this paper we give a natural description of the mapping between A and B and prove that the compression of A is *maximal*—that is, the compressed sequence B is itself incompressible—if and only if the martingale which induces the compression is *optimal* for A, i.e., A is Δ-random with respect to the measure corresponding to the martingale's betting strategy. A loose but intuitive interpretation of this result is that a bit stream can be maximally compressed (within a resource bound in Δ) if and only if it is Δ-random with respect to *some* computable measure (See Section 5).

A key technical fact used in the characterization described above, and one of the central results of this paper, is that resource-bounded randomness is preserved by a certain class of transformations on $\{0,1\}^\infty$, namely those implicit in the compressibility characterization of randomness. Our result extends to Δ-randomness an *invariance* property of algorithmic randomness discussed in [20] and in [6]; loosely, the binary expansion of a given real number x is Δ-random if and only if *every* representation of x as a binary sequence is Δ-random with respect to an underlying measure appropriate for the representation (Section 4).

We subsequently give a definition of compressibility with respect to nonuniform measures, which turns out to be a nontrivial extension of the original definition given in [2], and prove that the characterization of resource-bounded randomness still holds in the nonuniform case.

It is convenient to summarize our main results in the form of the following new characterizations of resource-bounded randomness. Here $A \in \{0,1\}^\infty$, Δ is a class of functions representing resource bounds, μ and ν represent measures which are Δ-computable, and d is a martingale with underlying measure ν and betting strategy determined by μ (see Section 2 for definitions and see Theorem 25 for a precise statement of hypotheses). The following are equivalent :

1. A is Δ-random with respect to μ.
2. The martingale d is optimal for A.
3. The martingale d induces a maximal compression of A.
4. A is Δ-incompressible with respect to μ.

In addition, in Section 3 we discuss the equivalence of measures and provide an extension to resource-bounded measure of the classical theorem of Kakutani on the equivalence of product measures, answering an open question in [1].

2 Preliminaries

2.1 Notation

Let $\mathbb{N}$ denote the natural numbers and let D denote the dyadic rationals (real numbers whose binary expansion is a finite string). A *string* is an element of $\{0,1\}^*$. The concatenation of strings x and y is denoted xy. For any string x, $|x|$

denotes the length of x, and λ is the unique string of length 0. If $x \in \{0,1\}^*$ and $j, k \in \mathbb{N}$ with $0 \leq j \leq k < |x|$, $x[k]$ is the kth bit (symbol) of x and $x[j..k]$ is the string consisting of the jth through kth bits of x. Note that the leftmost bit of x is $x[0]$; it is convenient to let $x[0..-1]$ denote the empty string. For $A \in \{0,1\}^\infty$, the notations $A[k]$ and $A[j..k]$ are defined analogously. For any $x, y \in \{0,1\}^*$, $x \sqsubseteq y$ means that if $x[k]$ is defined, then $y[k]$ is also defined and $x[k] = y[k]$; we say that x is an *initial segment*, or *prefix*, of y or that y is an *extension* of x. Likewise for $A \in \{0,1\}^\infty$, $x \sqsubseteq A$ means $x[k] = A[k]$ whenever bit $x[k]$ is defined. Strings x and y are said to be *incompatible*, or *disjoint*, if there is no string z which is an extension of both x and y, i.e., $x \not\sqsubseteq y$ and $y \not\sqsubseteq x$.

Fix a standard enumeration of $\{0,1\}^*$, $s_0 = \lambda, s_1 = 0, s_2 = 1, s_3 = 00, s_4 = 01, \ldots$. A *language* is a subset of $\{0,1\}^*$; a language A will be identified with its characteristic sequence $\chi_A \in \{0,1\}^\infty$, defined by $s_y \in A \iff \chi_A[y] = 1$ for $y \in \mathbb{N}$. We will consistently write A for χ_A. X^c denotes the complement of X in $\{0,1\}^\infty$. Strings used to represent partially defined languages (initial segments) will typically be represented by lower-case greek letters.

Throughout this paper the symbol Δ represents a class of functions, i.e., the resource bounds. We are generally interested in either rec, the class of all recursive functions, or p, the class of functions $f : \{0,1\}^* \to \{0,1\}^*$ such that $f(x)$ is computable in time polynomial in $|x|$, though the results are easily extended to many other resource bounds. A real-valued function f on $\{0,1\}^*$ is said to be Δ-*computable* if there is a function $\hat{f} : \mathbb{N} \times \{0,1\}^* \to D$ in Δ such that for all $n \in \mathbb{N}$ and $x \in \{0,1\}^*$, $|\hat{f}(n,x) - f(x)| \leq 2^{-n}$. A function f is *exactly* Δ-*computable* if f itself is in Δ and is dyadic-valued.

A string $\sigma \in \{0,1\}^*$ defines the subset $\text{Ext}(\sigma) = \{A \in \{0,1\}^\infty : \sigma \sqsubseteq A\}$ of $\{0,1\}^\infty$, called a *cylinder* or *interval*. Likewise if S is a subset of $\{0,1\}^*$, $\text{Ext}(S)$ denotes $\bigcup_{\sigma \in S} \text{Ext}(\sigma)$.

2.2 Measure

Definition 1. A *measure* is a function μ on $\{0,1\}^*$, taking values in $[0,1]$, such that $\mu(\lambda) = 1$ and for every $\sigma \in \{0,1\}^*$, $\mu(\sigma) = \mu(\sigma 0) + \mu(\sigma 1)$.

The function μ specifies a probability density on $\{0,1\}^\infty$, where $\mu(\sigma)$ represents the probability associated with the interval $\text{Ext}(\sigma)$. Standard results of measure theory (see [3]) show that such a function can always be extended uniquely to subsets $\mathcal{E} \subseteq \{0,1\}^\infty$ which are built up from intervals by some finite iteration of countable union and complementation operations (the *Borel* sets); we continue to write $\mu(\sigma)$ for $\mu(\text{Ext}(\sigma))$ and $\mu(S)$ for $\mu(\text{Ext}(S))$, where $S \subseteq \{0,1\}^*$. We may also regard a measure as a function on the unit interval $[0,1]$ via the usual correspondence between binary sequences and real numbers.

Let $\sigma \in \{0,1\}^*$, $|\sigma| = n$, and $b \in \{0,1\}$. If $\mu(\sigma) \neq 0$, the *bit probability*

$$\mu(\sigma b | \sigma) = \frac{\mu(\sigma b)}{\mu(\sigma)}$$

is defined, representing the conditional probability that bit n is equal to b, given the initial segment σ. By the multiplication rule for conditional probabilities, we always have

$$\mu(\sigma) = \prod_{i=0}^{n-1} \mu(\sigma[0..i] | \sigma[0..i-1])$$

when $\mu(\sigma) \neq 0$. Note that $\mu(\sigma 0 | \sigma) + \mu(\sigma 1 | \sigma) = 1$, and any function assigning a number $p_\sigma \in (0, 1)$ to each string σ uniquely determines via the multiplication rule a measure μ whose bit probabilities are $\mu(\sigma 0 | \sigma) = 1 - p_\sigma$ and $\mu(\sigma 1 | \sigma) = p_\sigma$.

Most of our results will apply to measures for which all the bit probabilities are either positive or are bounded away from zero. The terminology "strongly positive" was suggested in [1].

Definition 2. A measure μ is *positive* if $\mu(\sigma) > 0$ for every string σ. A measure μ is *strongly positive* if there is a constant $\delta > 0$ such that for every string σ and $b \in \{0, 1\}$, $\delta \leq \mu(\sigma b | \sigma) \leq 1 - \delta$.

If the bit probabilities are *independent*, that is, $\mu(\sigma b | \sigma)$ depends only on the position $n = |\sigma|$ but not on σ itself, the measure μ is then referred to as a *product measure*; the measure is determined by a sequence of pairs $\{(1 - p_n, p_n)\}_{n=0}^{\infty}$, i.e., $\mu(\sigma 1 | \sigma) = p_n$ for all σ of length n. The sequence $\{(1 - p_n, p_n)\}$ may be viewed as sequence of coins, fixed in advance, such that the nth coin has probability p_n of coming up heads. If all the coins are fair ($p_n = \frac{1}{2}$), then μ is the usual Lebesgue measure on the Borel subsets of $\{0, 1\}^{\infty}$. We use the symbol λ for Lebesgue measure (distinguished from the empty string, hopefully, in context).

If μ is not a product measure, μ is said to contain *dependencies*. We may still use the intuitive picture of a sequences of coin tosses, but with the following difference. Instead of the sequence of coins being fixed in advance of the experiment, there is a daemon which examines the outcomes σ of the first n tosses and then selects a coin whose probability of coming up heads is $\mu(\sigma 1 | \sigma)$. Measures with dependencies are probably the appropriate models for imperfect, physical sources of randomness where it generally cannot be assumed that the bit probabilities are independent. Examples of such measures include the "adversary sources" of [13], where in analyzing the behavior of probabilistic algorithms using such a source the authors assume the source to be adversarial, that is, the daemon determining the function μ is aware of the probabilistic algorithm being used and attempts to fix the bias of successive coins, within some bounds $[\delta, 1 - \delta]$, in such a way as to make the algorithm fail.

2.3 Martingales

Definition 3. Let ν be a positive measure. A ν-*martingale* is a function d on $\{0, 1\}^*$ taking values in $[0, \infty)$ such that

$$d(\sigma) = \nu(\sigma 0 | \sigma) d(\sigma 0) + \nu(\sigma 1 | \sigma) d(\sigma 1). \tag{1}$$

A martingale d *succeeds* on a sequence $A \in \{0,1\}^\infty$ if

$$\limsup_{n \to \infty} d(A[0..n]) = \infty.$$

A martingale may be regarded as a strategy for betting on successive bits of an infinite sequence. The value $d(\sigma)$ represents the gambler's accumulated capital after sequence of outcomes σ. The equality (1) asserts that the game is *fair*, that is, at each node σ the (conditional) expected value of the capital after the next bit always equal to its present value.

It follows from the standard result below, known as *Kolmogorov's inequality for martingales*, that a set $X \subseteq \{0,1\}^\infty$ has *measure zero* with respect to measure ν if and only if there is a ν-martingale succeeding on every $A \in X$: that is, a martingale may be viewed as an *orderly* demonstration that a set X has measure zero. It is this fact which makes martingales useful for defining measure with resource bounds.

Lemma 4. *Let d be a ν-martingale and t a positive real number; then*

$$\nu\{\sigma \in \{0,1\}^* : d(\sigma) > t\} < \frac{d(\lambda)}{t}.$$

2.4 Resource-bounded Measure

Resource-bounded measure theory, as developed by Lutz, is a form of effective or constructive measure theory which provides a means of defining the measure or probability of sets of languages within many standard complexity classes, allowing for the first time a *quantitative* analysis of such well-studied questions as, for example, whether P is a proper subclass of NP. In addition, a definition of measure within a complexity class provides a means of defining the "random" languages for the class. See [10], [8], or [11] for a more thorough introduction.

Definition 5. Let Δ be a class of functions, and let ν be a Δ-computable measure. A class $X \subseteq \{0,1\}^\infty$ has Δ-*measure zero with respect to* ν, written $\nu_\Delta(X) = 0$, if there is a Δ-computable ν-martingale which succeeds on every A in X. X has Δ-*measure one w.r.t.* ν if $\nu_\Delta(X^c) = 0$. A sequence $A \in \{0,1\}^\infty$ is Δ-*random w.r.t* ν if there is no Δ-computable ν-martingale which succeeds on A.

Lutz has proved a number of results showing that Δ-measure behaves like a measure within an appropriate corresponding complexity class; for example, p-measure is appropriate as a measure in the class E $= \text{DTIME}(2^{\text{linear}})$.

3 The Decomposition of a Martingale Using Likelihood Ratios

A martingale d may be fruitfully viewed as composed of two parts, which correspond respectively to the *strategy* used by the gambler and to *odds* paid on her wins. At each node σ, the gambler selects some proportion p_σ of her capital $d(\sigma)$ to bet on $\sigma 1$ and the remaining proportion $1 - p_\sigma$ to bet on $\sigma 0$. The *strategy* is the unique measure μ determined by the bit probabilities $\mu(\sigma 1|\sigma) = p_\sigma$.

It is the underlying measure ν which determines the odds. Recall that in a fair game, if a bet of amount B is placed on an event $\mathcal{E}$, and $\mathcal{E}$ occurs, the gambler receives her original B plus $\frac{N}{D}B$ more, where the ratio N/D is equal to the probability of **not** $\mathcal{E}$ divided by the probability of $\mathcal{E}$, the so-called odds against $\mathcal{E}$, usually expressed as "N to D" with N, D integers. Let q_σ denote the bit probability $\nu(\sigma 1|\sigma)$ and let $p_\sigma = \mu(\sigma 1|\sigma)$, the proportion of $d(\sigma)$ the gambler bets on $\sigma 1$. The odds paid on the occurrence of a "1" are $(1 - q_\sigma)/q_\sigma$, so we can compute the capital at $\sigma 1$ as

$$d(\sigma 1) = p_\sigma d(\sigma)\left(1 + \frac{1 - q_\sigma}{q_\sigma}\right) = \frac{p_\sigma}{q_\sigma} \cdot d(\sigma),$$

and similarly, $\quad d(\sigma 0) = \dfrac{1 - p_\sigma}{1 - q_\sigma} \cdot d(\sigma).$

Using the multiplication rule inductively we have, for all σ,

$$d(\sigma) = \frac{\mu(\sigma)}{\nu(\sigma)} \cdot d(\lambda). \tag{2}$$

We may refer to d as the *ν-martingale with strategy μ*. The coefficient $\mu(\sigma)/\nu(\sigma)$ is known as a *likelihood ratio* (see [16]). In the next two subsections we investigate how the relationship between ν and μ affects the success of the martingale.

Most of our results will apply only to martingales for which the underlying measure ν is strongly positive, and in many cases an assumption is required that the measure μ determining the strategy be strongly positive as well. The next lemma shows that very little generality is lost by the latter assumption. In addition we will need the fact shown below that "limsup" can be replaced by "lim" in Definition 3.

Lemma 6. *Let d be a Δ-computable ν-martingale, where ν is a strongly positive, Δ-computable measure. There exists a strongly positive, Δ-computable measure μ such that, if $\tilde{d}$ denotes the ν-martingale with strategy μ, then for every A on which d succeeds, $\lim_{n\to\infty} \tilde{d}(A[0..n]) = \infty$.*

3.1 Optimal Martingales

Here we present a simple application of the decomposition via likelihood ratios given above. This notion of "optimality" will also be used in the proof of Theorem 16. Note first that it is clear from the definitions that if a sequence A is Δ-random w.r.t. a measure ν, then for any measure μ the ratio $\mu(A[0..n])/\nu(A[0..n])$ is bounded, since no ν-martingale can succeed on A. On the other hand, if A is random with respect to μ but not ν, the martingale $d(\sigma) = \mu(\sigma)/\nu(\sigma)$ not only succeeds on A, but does so at an asymptotically optimal rate. In fact, the converse holds as well, providing the first of the characterizations of Δ-randomness promised in the introduction.

Definition 7. Fix a measure ν; a ν-martingale d is *optimal* for $A \in \{0,1\}^\infty$ if for every other ν-martingale $\tilde{d}$, there is a constant C such that for all n, $\tilde{d}(A[0..n]) < C \cdot d(A[0..n])$.

Theorem 8. *Let μ and ν be strongly positive, Δ-computable measures, let d be the ν-martingale with strategy μ, and let $A \in \{0,1\}^\infty$. Then A is Δ-random with respect to μ if and only if d is optimal for A.*

3.2 Equivalence of Measures

It is clear from the form of (2) that if the strategy μ of a martingale is nearly the same as the underlying measure ν, the martingale will not succeed on any sequence A. In this section we characterize just how different μ and ν must be in order for the martingale to succeed. It turns out that there are two useful forms of equivalence we might consider, which we refer to here as a "strong" form and a "weak" form. What we refer to as strongly equivalent measures are those which are completely interchangeable in the construction of martingales as in (2).

Definition 9. Let μ and ν be positive measures and let $A \in \{0,1\}^\infty$. We say that μ and ν are *strongly equivalent* if there are constants $C > c > 0$ such that for all $A \in \{0,1\}^\infty$ and all n, $c < \mu(A[0..n])/\nu(A[0..n]) < C$.

On the other hand, what we call "weak" equivalence corresponds to the classical notion of *absolute continuity*, i.e., the measure zero sets for both measures are the same.

Definition 10. Let μ and ν be Δ-computable measures. We say that μ and ν are *weakly equivalent* if for every $X \subseteq \{0,1\}^\infty$, $\mu_\Delta(X) = 0$ if and only if $\nu_\Delta(X) = 0$.

There is a simple and useful characterization of strong equivalence in terms of bit probabilities. The following is found in [1].

Lemma 11. *Let μ and ν be strongly positive measures; for $A \in \{0,1\}^\infty$ and $i \in \mathbb{N}$ let $u_i = \mu(A[0..i]|A[0..i-1])$ and $v_i = \nu(A[0..i]|A[0..i-1])$. Then μ and ν are strongly equivalent if and only if for every $A \in \{0,1\}^\infty$, $\sum_{i=0}^\infty |u_i - v_i| < \infty$.*

A fact about strongly equivalent measures which we will use repeatedly is the following "exact computation lemma" for measures, which is found in [1].

Lemma 12. *Let μ be a strongly positive, Δ-computable measure. Then there is an exactly Δ-computable measure ν which is strongly equivalent to μ.*

We now turn to the characterization of weak equivalence. A well-known theorem of Kakutani [5] on infinite product measures characterizes the (weak) equivalence of two measures in a manner similar to Lemma 11, but in terms of the weaker condition that the *squares* of the differences between bit probabilities are summable. It is also known that Kakutani's theorem holds for constructive measure in the sense of Martin-Löf (see [14]). What we show below is that Kakutani's theorem holds for resource-bounded measure, answering an open question in [1]. Just as for Kakutani's original result, the converse of Theorem 13 also holds if μ and ν are *product* measures.

Theorem 13. *Let μ and ν be strongly positive, Δ-computable measures, and for $A \in \{0,1\}^\infty$, $i \in \mathbb{N}$, let $u_i = \mu(A[0..i]|A[0..i-1])$ and $v_i = \nu(A[0..i]|A[0..i-1])$. Suppose that for each $A \in \{0,1\}^\infty$, $\sum_{i=0}^\infty (u_i - v_i)^2 < \infty$. Then μ and ν are weakly equivalent.*

4 An Invariance Property of Δ-randomness

In this section we prove a fundamental technical result which shows that Δ-randomness is invariant under certain kinds of transformations on $\{0,1\}^\infty$. This result is of independent interest since it extends to Δ-randomness some known results on invariance properties of algorithmic randomness; however, we are primarily concerned here because this result is the key to our understanding of the notion of compressibility in [2] (see Definition 17) and the subsequent characterization of Δ-randomness in terms of "maximal compressions."

We begin with a simple idea, namely, the representation of a real number with respect to a given measure. While this topic may at first appear to be a digression, it is surely the most intuitive way to introduce the kinds of transformations we need to discuss. Some of the results here come from the detailed treatment in [6], and as implied above this kind of transformation appears implicitly in [2]; however, the idea can really be traced back to the "isomorphism theorem" for Lebesgue measure (see [4]). The same transformation can also be found in data compression algorithms based on arithmetic coding; see [19].

The following is a natural way, given a measure μ, to associate with a given string σ a subinterval of $[0, 1]$ of width $\mu(\sigma)$.

Definition 14. Let μ be a measure and $\sigma \in \{0,1\}^*$. Let S be a maximal, disjoint set of strings which lexicographically precede σ (e.g., the lexicographic predecessors of length $|\sigma|$). The *basic μ-interval* $(\sigma)_\mu$ is the interval $[p, q] \subseteq [0, 1]$ defined by $p = \sum_{\tau \in S} \mu(\tau)$, $q = p + \mu(\sigma)$. More generally, for any subinterval $[x, y] \subseteq [0, 1]$, let $([x, y])_\mu = [\mu([0, x]), \mu([0, y])]$.

For example, $(\sigma)_\lambda$ is what we would normally call a dyadic interval (remember that λ here denotes Lebesgue measure). It is not difficult to see that in the case of Lebesgue measure, the definition below gives the usual interpretation of the binary expansion of a real number in $[0, 1]$.

Definition 15. Let μ be a strongly positive measure and $A \in \{0,1\}^\infty$. The *real number μ-represented by A*, which we denote $(A)_\mu$, is the unique real number x such that

$$x \in \bigcap_i (A[0..i])_\mu.$$

The main result of this section asserts that the transformation between representations preserves Δ-randomness.

Theorem 16. *Let μ and ν be strongly positive, Δ-computable measures. Let A and B be sequences such that $(A)_\mu = (B)_\nu$. Then A is Δ-random w.r.t. μ if and only if B is Δ-random w.r.t. ν.*

Theorem 16 may be interpreted as an assertion that Δ-randomness is an invariant property of real numbers, that is, a property which is independent of the scheme used to represent real numbers as binary sequences. The analog of Theorem 16 for algorithmic randomness is well-known; Theorem 16 is unusual in that its proof does not seem to depend on computational properties of the transformation between A and B. Note that given the ν-representation B of some real number x, the μ-representation A of x can be computed as follows: Having determined $A[0..i]$, choose $b = A[i+1]$ so that some interval $(B[0..j])_\nu$ is completely contained in $(A[0..i]b)_\mu$. Although it is not obvious, in most cases of interest it turns out that $A \leq_{tt} B$, and so the analog of Theorem 16 for algorithmic randomness is closely related to the fact that algorithmic randomness is always preserved by tt-reductions; see [20] or [6]. It is unknown whether Δ-randomness is always preserved by tt-reductions, even in the case $\Delta = \text{rec}$, although Breutzmann and Lutz [1] have shown that Δ-randomness is preserved by a certain restricted class of tt-reductions.

5 Compressibility

Buhrman and Longpré [2] gave the following definition of "compressibility" and the subsequent characterization of Δ-measure zero.

Definition 17. $A \in \{0,1\}^\infty$ is f-compressible if there exists a sequence $B \in \{0,1\}^\infty$ and algorithms C and D such that

1. The algorithm C ("compression"), given an initial segment $A[0..i]$, produces in $f(i)$ steps a finite number of strings, one of which is an initial segment $B[0..j]$ such that D, on input $B[0..j]$, produces a prefix of A which properly extends $A[0..i]$.
2. The algorithm D ("decompression"), given $B[0..j]$, produces in $f(i+j)$ steps a prefix $A[0..i]$; moreover, the value $i - j$ is unbounded.

We say A is Δ-compressible if A is f-compressible for some f in Δ.

Theorem 18. *A set $X \subseteq \{0,1\}^\infty$ has Δ-measure zero if and only if for some f in Δ, every A in X is f-compressible.*

In the proof that measure zero implies compressibility, given in [2], a particular martingale succeeding on a sequence A is used to define the "compressed" sequence B. It is not difficult to see from the proof in [2] that the sequence B is precisely the standard representation of the real number x whose μ-representation is the original sequence A, where μ is the betting strategy of the martingale. (A sketch of the argument is given below following Lemma 19.)

We have seen that if A is Δ-random with respect to μ, then the martingale with strategy μ is optimal for A. Intuitively it would make sense that an "optimal" martingale should yield a "maximal" compression, that is, the compressed sequence should itself be incompressible. This is the idea behind the characterization given in Theorem 22 below.

It is worthwhile to sketch the arguments leading to Theorem 22. Although the result will be generalized by Theorem 24, the ideas can be more easily seen in this simpler context, where the underlying measure for all martingales is λ (Lebesgue measure) and so we can use Definition 17 without modification. The first step is to show that the optimal martingale of Theorem 8 does indeed yield a compression according to Definition 17.

Lemma 19. *Let μ be a strongly positive, exactly Δ-computable measure and let d be the λ-martingale with strategy μ. Suppose that $\lim_n d(A[0..n]) = \infty$; then d defines a compression of A in the sense of Definition 17, where the compressed sequence B is the standard representation of the real number $(A)_\mu$.*

The proof of Lemma 19 is not essentially different from the reasoning originally given in [2], but we need to show that some of the restrictions on martingales imposed in [2] can be relaxed. The following technical lemma (which is also used in the proofs of Theorems 16 and 24) will absorb many of the detailed restrictions on the martingales used in [2].

Lemma 20. *Let μ and ν be strongly positive, Δ-computable measures, and let $\delta > 0$ such that all bit probabilities for μ and ν lie in $[\delta, 1 - \delta]$. Let A and B be the μ- and ν-representations, respectively, of some real number; i.e., $(A)_\mu = (B)_\nu$. For each j, let $A[0..i_j]$ denote the largest initial segment of A for which $(B[0..j])_\nu \subseteq (A[0..i_j])_\mu$. Then for infinitely many j, $\nu(B[0..j]) \geq \delta^2 \mu(A[0..i_j])$.*

We will also refer to the following simple fact.

Lemma 21. *Let x and y be dyadic rational numbers with representations at most n bits long. Then $[x, y]$ can be exactly covered, in a unique way, with fewer than $2n$ nonoverlapping dyadic intervals of maximal width, each of which has a representation of at most n bits.*

Sketch of proof of Lemma 19. The idea is that each initial segment $A[0..i]$ determines a subinterval $(A[0..i])_\mu$ of $[0, 1]$, and the sequence B is the standard binary representation of the real number $(A)_\mu = \bigcap_i (A[0..i])_\mu$. The "decompression" algorithm is essentially the reduction from B to A described following Theorem 16; given an initial segment $B[0..j]$, find the longest string σ such that $(B[0..j])_\lambda \subseteq (\sigma)_\mu$. It necessarily follows that $\sigma = A[0..i]$ is the longest initial segment of A such that $(B[0..j])_\lambda \subseteq (A[0..i])_\mu$. Intuitively the intervals $(A[0..i])_\mu$ and $(B[0..j])_\lambda$ should represent about the same amount of information, i.e., the width of $(A[0..i])_\mu$ should be of the same order as the width of $(B[0..j])_\lambda$, or about 2^{-j}, so we would expect

$$\frac{2^{-j}}{2^{-i}} \approx \frac{\mu(A[0..i])}{\lambda(A[0..i])}. \tag{3}$$

Since the right-hand side is unbounded, it would then follow that $i - j$ is unbounded.

Of course, (3) is not literally true, but by Lemma 20 there are infinitely many $j \in \mathbb{N}$ such that the decompression algorithm produces an initial segment $A[0..i_j]$ satisfying $\lambda(B[0..j]) \geq \delta^2 \mu(A[0..i_j])$. Since $\lim_n d(A[0..n]) = \infty$, this is sufficient to conclude that $i - j$ is unbounded as required by Definition 17.

The "compression" algorithm is obtained by applying Lemma 21 to each of the two subintervals $(A[0..i]0)_\mu$ and $(A[0..i]1)_\mu$, to produce the list of strings required by the definition. One of these strings must be an initial segment $B[0..j]$; since $(B[0..j])_\lambda$ is completely contained in either $(A[0..i]0)_\mu$ or $(A[0..i]1)_\mu$, the decompression algorithm produces a proper extension of $A[0..i]$. □

The remaining key ingredient in the proof of Theorem 22 is the invariance property given by Theorem 16. Suppose A is Δ-compressible and is Δ-random with respect to a strongly positive, Δ-computable measure μ. By Lemma 12 we can assume μ is exactly Δ-computable, and by Theorem 8 we know that the martingale $d(\sigma) = \mu(\sigma)/\lambda(\sigma)$ is optimal for A, and hence it follows from Lemma 6 that $\lim_n d(A[0..n]) = \infty$. Thus this optimal d satisfies the hypotheses of Lemma 19. Since A is Δ-random w.r.t. μ, Theorem 16 shows that the compressed sequence B is Δ-random (with respect to λ), i.e., A is "maximally compressed" by d. Conversely, if A is not Δ-random w.r.t. μ, the compressed sequence B is not Δ-random w.r.t. λ. Thus we have the following characterization:

Theorem 22. *Let μ be a strongly positive, Δ-computable measure, and suppose $A \in \{0,1\}^\infty$ is Δ-compressible. Let d be the λ-martingale with strategy μ. Then A is Δ-random w.r.t. μ if and only if d induces a maximal compression of A.*

In the next section we will be able to replace the references to Lebesgue measure λ in the above characterization with an arbitrary measure ν.

6   Compressibility with Respect to Nonuniform Measures

It is not entirely obvious how to extend Definition 17 to an arbitrary measure ν, that is, to do so in such a way as to continue to be able to characterize the notion of Δ-measure zero with respect to ν. The "easy" part is to replace the condition "$i - j$ is unbounded" in Definition 17 with the condition (4) below; i.e., see the left-hand side of (3). The peculiar feature of Lebesgue measure which is used in an essential way in Theorem 18 is that every dyadic number x in $[0, 1]$ occurs as the endpoint of some basic dyadic interval $(\sigma)_\lambda$, where $|\sigma|$ is no longer than the representation of x; thus, any interval $[x, y]$ with dyadic endpoints can be *exactly* and *efficiently* covered with basic λ-intervals $(\sigma)_\lambda$ (see Lemma 21).

Very loosely, the proof that measure zero implies compressibility uses the following idea. Given an exact martingale d with strategy μ succeeding on A, the compression algorithm takes an initial segment $\sigma = A[0..i]$, finds the end-points of the intervals $(\sigma 0)_\mu$ and $(\sigma 1)_\mu$, and produces a list of strings τ (the "candidates") representing a set of dyadic intervals $(\tau)_\lambda$ which exactly cover $(\sigma 0)_\mu$ and $(\sigma 1)_\mu$. These candidates can be divided into two groups G_0 and G_1, those which "decompress" into extensions of $\sigma 0$ and $\sigma 1$, respectively. Then in the other direction of the proof ("compressibility implies measure zero") it is the relative measures of G_0 and G_1 which are used to define the values $d(\sigma 0)$ and $d(\sigma 1)$ of a martingale d. The difficulties arise from the fact that since $(\sigma 0)_\mu$ and $(\sigma 1)_\mu$ cannot be *exactly* covered by basic ν-intervals, the sets G_0 and G_1 must overlap. The condition (5) below is imposed to ensure that the amount of overlap is controlled. These and related issues are discussed at greater length in the full version of the paper.

Definition 23. Let $A \in \{0,1\}^\infty$, let ν be a Δ-computable measure, and let f be a function in Δ. A is *f-compressible with respect to ν* if there exists $B \in \{0,1\}^\infty$, algorithms C and D with running time bounded by f, and a summable sequence $\{\epsilon_i\}$, $0 < \epsilon_i < 1$, satisfying the following conditions:

1. The function C ("compression") takes a string σ as input and produces a pair $C(\sigma) = \{G_0(\sigma), G_1(\sigma)\}$ of finite sets of strings, called *candidates*, such that each string τ appearing in $G_0(\sigma b)$ or $G_1(\sigma b)$ extends some τ' in $G_b(\sigma)$. For every initial segment $\sigma b \sqsubseteq A$, there exists a string $\tau \in G_b(\sigma)$ such that $\tau \sqsubseteq B$ and $\sigma b \sqsubseteq D(\tau) \sqsubseteq A$.

2. The function D ("decompression") takes strings to strings; whenever any string $\tau \sqsubseteq B$ appears as a candidate in $G_b(\sigma)$ for some $\sigma b \sqsubseteq A$, then $\sigma b \sqsubseteq D(\tau) \sqsubseteq A$. Moreover, given any constant k there is an initial segment $A[0..i] = \sigma b$ such that $G_b(\sigma)$ contains a candidate $B[0..j]$ for which

$$\frac{\nu(B[0..j])}{\nu(A[0..i])} > k. \tag{4}$$

3. For every $\sigma \sqsubseteq A$, $i = |\sigma|$,

$$\nu(G_0(\sigma) \cup G_1(\sigma)) \geq \nu(G_0(\sigma)) + \nu(G_1(\sigma)) - \epsilon_i \nu(G_1(\sigma)). \tag{5}$$

We may say that A is *Δ-incompressible* if A is not f-compressible for any $f \in \Delta$.

The following theorem then provides the analog to Theorem 18.

Theorem 24. *Let ν be a strongly positive, Δ-computable measure, and let $X \subseteq \{0,1\}^\infty$. Then X has Δ-measure zero with respect to ν if and only if there is a function $f \in \Delta$ such that every $A \in X$ is f-compressible with respect to ν.*

In the course of proving Theorem 24 we show that the analog of Lemma 19 holds for compressibility with respect to an arbitrary strongly positive measure ν, and thus the analog of Theorem 22 holds for compressibility with respect to ν via similar reasoning.

We now have all the characterizations promised:

Theorem 25. *Let μ and ν be strongly positive, Δ-computable measures, and let d be the ν-martingale with strategy μ. Suppose $A \in \{0,1\}^\infty$ is not Δ-random w.r.t. ν. The following are equivalent.*

1. *A is Δ-random with respect to μ.*
2. *d is optimal for A.*
3. *d induces a maximal compression of A with respect to ν.*
4. *A is Δ-incompressible with respect to μ.*

7 Conclusion

We suggest that the main contributions of this paper are the invariance property of Theorem 16 and its subsequent application in the characterization of randomness given in Theorem 22. We showed in Theorem 16 that resource-bounded randomness is preserved by an important class of transformations on $\{0,1\}^\infty$. Such transformations are most simply described as mappings between different representations of a given real number, and they can also be seen in data compression algorithms based on arithmetic coding, but most importantly, in Buhrman and Longpré's characterization of Δ-randomness as incompressibility the canonical compression scheme induced by a martingale is seen to be an instance of such a transformation. After solving a number of technical problems we were then able to conclude in Theorem 22 that the compressed sequence is itself Δ-random with respect to the uniform distribution—that is, incompressible—if and only if the original sequence is random with respect to the measure used as the martingale's betting strategy. Thus, if we restrict our attention to strongly positive measures, Δ-randomness is an *intrinsic* property of a sequence: If the sequence is Δ-random with respect to some (Δ-computable) measure, then within the resource class Δ a canonical transformation will yield a uniformly Δ-random sequence by making the representation as efficient as possible (maximally compressed); if the sequence is not Δ-random with respect to any measure, there is no way within a bound in Δ to maximally compress the sequence or to transform it into a random one.

In the course of developing the results above we also proved several facts of independent interest. We showed, using the idea of likelihood ratios, that the optimal betting strategy for a martingale to use on a given sequence A is to bet a proportion of its capital which precisely matches the probability distribution of A. We also investigated the notion of equivalence of measures and proved that two measures are equivalent—that is, the measure zero sets, and therefore the random sequences, are the same for both—if and only if the differences between corresponding bit probabilities are square-summable, thus extending to resource-bounded measure a classical theorem of Kakutani.

Finally we investigated the issues involved in extending the compressibility characterization of Buhrman and Longpré to nonuniform measures. We proposed a definition of compressibility in the spirit of the original, and proved that a sequence is Δ-incompressible with respect to a measure ν if and only if it is Δ-random with respect to ν.

211

References

1. J.M. Breutzmann and J. H. Lutz. Equivalence of measures of complexity classes. 1996. To appear.
2. H. Buhrman and L. Longpré. Compressibility and resource bounded measure. 1995 STACS.
3. W. Feller. *An Introduction to Probability Theory and its Applications.* Volume 2, John Wiley and Sons, Inc., 1971.
4. P.R. Halmos. *Measure Theory.* Springer-Verlag, 1974.
5. S. Kakutani. On the equivalence of infinite product measures. *Annals of Mathematics*, 49:214–224, 1948.
6. S. M. Kautz. *Degrees of Random Sets.* PhD thesis, Cornell University, 1991.
7. M. Li and P. Vitányi. *An Introduction to Kolmogorov Complexity and Its Applications.* Springer-Verlag, 1993.
8. J. H. Lutz. The quantitative structure of exponential time. In *Proceedings of the Eighth Annual Structure in Complexity Theory Conference*, pages 158–175, 1993. Updated version to appear in L. A. Hemaspaandra and A. L. Selman (eds.), *Complexity Theory Retrospective II*, Springer-Verlag, 1996.
9. J. H. Lutz and E. Mayordomo. Genericity, measure, and inseparable pairs. 1996. In preparation.
10. J.H. Lutz. Almost everywhere high nonuniform complexity. *Journal of Computer and System Sciences*, 44:220–258, 1992.
11. E. Mayordomo. *Contributions to the Study of Resource-Bounded Measure.* PhD thesis, Universitat Politècnica de Catalunya, 1994.
12. Noam Nisan. Extracting randomness: how and why. A survey. In *Proceedings of the 11th IEEE Conference on Computational Complexity*, 1996.
13. M. Santha and U.V. Vazirani. Generating quasi-random sequences from slightly-random sources. In *Proc. 25th Ann. Symp. on the Theory of Computing*, 1984.
14. A. Kh. Shen´. Algorithmic complexity and randomness: recent developments. *Theory Probab. Appl.*, 37(3):92–97, 1993.
15. A. Kh. Shen´. On relations between different algorithmic definitions of randomness. *Soviet Math. Dokl.*, 38(2):316–319, 1989.
16. M. van Lambalgen. *Random Sequences.* PhD thesis, University of Amsterdam, 1987.
17. M. van Lambalgen. Von Mises' definition of random sequences reconsidered. *Journal of Symbolic Logic*, 52(3):725–755, 1987.
18. U.V. Vazirani and V.V. Vazirani. Random polynomial time is equal to slightly-random polynomial time. In *Proceedings of the 26th IEEE Symposium on Foundations of Computer Science*, 1985.
19. I. H. Witten, R. M. Neal, and J. G. Cleary. Arithmetic coding for data compression. *Communications of the Association for Computing Machinery*, 30:520–540, 1987.
20. A.K. Zvonkin and L.A. Levin. The complexity of finite objects and the development of the concepts of information and randomness by means of the theory of algorithms. *Russian Mathematical Surveys*, 25:83–123, 1970.

Randomness, Stochasticity and Approximations*

Yongge Wang**

Department of Computer Science, The University of Auckland, Private Bag 92019, Auckland, New Zealand (wang@cs.auckland.ac.nz)

Abstract. Polynomial time unsafe approximations for intractable sets were introduced by Meyer and Paterson [9] and Yesha [19] respectively. The question of which sets have optimal unsafe approximations has been investigated extensively, see, e.g., [1, 5, 15, 16]. Recently, Wang [15, 16] showed that polynomial time random sets are neither optimally unsafe approximable nor Δ-levelable. In this paper, we will show that: (1) There exists a polynomial time stochastic set in $\mathbf{E}_2$ which has an optimal unsafe approximation. (2). There exists a polynomial time stochastic set in $\mathbf{E}_2$ which is Δ-levelable. The above two results answer a question asked by Ambos-Spies and Lutz et al. [3]: Which kind of natural complexity property can be characterized by p-randomness but not by p-stochasticity? Our above results also extend Ville's [13] historical result. The proof of our first result shows that, for Ville's stochastic sequence, we can find an optimal betting strategy (prediction function) such that we will never lose our own money (except the money we have earned), that is to say, if at the beginning we have only one dollar and we always bet one dollar that the next selected bit is 1, then we always have enough money to bet on the next bit. Our second result shows that there is a stochastic sequence for which there is a betting strategy such that we will never lose our own money (except the money we have earned), but there is no such kind of optimal betting strategy. That is to say, for any such kind of betting strategy, we can find another betting strategy which could be used to make money more quickly.

1 Introduction

Random sequences were first introduced by von Mises [10] as a foundation for probability theory. Von Mises thought that random sequences were a type of disordered sequences, called "Kollektivs". The two features characterizing a Kollektiv are: the existence of limiting relative frequencies within the sequence and the invariance of these limits under the operation of an "admissible place selection

---

* The work reported here is supported by the Centre for Discrete Mathematics and Theoretical Computer Science (CDMTCS) at the University of Auckland, and by the postdoctoral fellowship (supervisor: Professor Cristian Calude) of the University of Auckland.

** I would also like to thank Professor Harald Ganzinger who supported my stay at Max-Planck-Institut für Informatik (Saarbrücken, Germany) where part of this work was done.

rule". Here an admissible place selection rule is a procedure for selecting a subsequence of a given sequence ξ in such a way that the decision to select a bit $\xi[n]$ does not depend on the value of $\xi[n]$. But von Mises' definition of an "admissible place selection rule" is not rigorous according to modern mathematics. After von Mises introduced the concept of "Kollektivs", the first question raised was whether this concept is consistent. Wald [14] answered this question affirmatively by showing that, for each countable set of admissible place selection rules, the corresponding set of "Kollektivs" has Lebesgue measure 1. The second question raised was whether all "Kollektivs" satisfy the standard statistical laws. For a negative answer to this question, Ville [13] constructed a counterexample in 1939. He showed that, for each countable set of admissible place selection rules, there exists a "Kollektiv" which does not satisfy the law of the iterated logarithm. The example of Ville defeated the plan of von Mises to develop probability theory based on "Kollektivs", that is to say, to give an axiomatization of probability theory with "random sequences" (i.e., "Kollektivs") as a primitive term. Later, admissible place selection rules were further developed by Tornier, Wald, Church, Kolmogorov, Loveland and others. This approach of von Mises to define random sequences is now known as the "stochastic approach".

A completely different approach to the definition of random sequences was proposed by Martin-Löf [8]. He developed a quantitative (measure-theoretic) approach to the notion of random sequences. This approach is free from those difficulties connected with the frequency approach of von Mises. Later, Schnorr [11] used the martingale concept to give a uniform description of various notions of randomness. In particular, he gave a characterization of Martin-Löf's randomness concept in these terms.

Using martingales concepts, Schnorr [11] introduced resource bounded randomness concepts, and later Lutz [7] introduced a kind of resource bounded measure theory. Resource bounded version of stochasticity concepts were also introduced by several authors, see, e.g., Wilber [18], Ko [6] and Ambos-Spies et al. [2].

The notion of unsafe approximations was introduced by Yesha in [19]: An unsafe approximation algorithm for a set A is just a standard polynomial time bounded deterministic Turing machine M with outputs 1 and 0. Duris and Rolim [5] further investigated unsafe approximations and introduced a levelability concept, Δ-levelability, which implies the nonexistence of optimal polynomial time unsafe approximations. They showed that complete sets for $\mathbf{E}$ are Δ-levelable and there exists an intractable set in $\mathbf{E}$ which has an optimal safe approximation but no optimal unsafe approximation. But they did not succeed to produce an intractable set with optimal unsafe approximations. Ambos-Spies [1] defined a concept of weak Δ-levelability and showed that there exists an intractable set in $\mathbf{E}$ which is not weakly Δ-levelable (hence it has an optimal unsafe approximation). In [15, 16], Wang extended Ambos-Spies's results by showing that both the class of Δ-levelable sets and the class of sets which have optimal polynomial time unsafe approximations have p-measure 0. Wang's results show that Δ-levelable sets and optimally approximable sets could not be p-random. However, in this paper, we will show the following results.

- There is a p-stochastic set in $\mathbf{E}_2$ which has an optimal unsafe approximation.
- There is a p-stochastic set in $\mathbf{E}_2$ which is Δ-levelable.

Note that our above results extend Ville's [13] historical result. Ville's result says that: For every countable set of admissible place selection rules, we can construct a stochastic sequence ξ which has more 1s than 0s in its initial segments. As we will show in Theorem 20, for this stochastic sequence ξ, the prediction function $f(x) = 1$ will be the optimal prediction strategy since, for every other prediction function g, there is a $k \in N$ such that $\|\{i < n : g(\xi[0..i-1]) = \xi[i]\}\| \leq \|\{i < n : f(\xi[0..i-1]) = \xi[i]\}\| + k$ for almost all $n \in N$. Our second result (Lemma 21 and Theorem 22) says that: For every countable set of admissible place selection rules, we can construct a stochastic sequence ξ such that there is no optimal prediction strategy for this sequence. That is to say, for every prediction function f, there is another prediction function g and an unbounded nondecreasing function $r(n)$ such that $\|\{i < n : g(\xi[0..i-1]) = \xi[i]\}\| \geq \|\{i < n : f(\xi[0..i-1]) = \xi[i]\}\| + r(n)$ for almost all $n \in N$. We will prove our results for the resource bounded case only, but all of these results hold for the classical case also.

The outline of the paper is as follows. In section 3 we review the relations between the concept of resource bounded randomness and the concept of polynomial time unsafe approximations. In section 4 we establish the relations between the concept of resource bounded stochasticity and the concept of polynomial time unsafe approximations.

2 Definitions

N and $Q(Q^+)$ are the set of natural numbers and the set of (nonnegative) rational numbers, respectively. $\Sigma = \{0, 1\}$ is the binary alphabet, Σ^* is the set of (finite) binary strings, Σ^n is the set of binary strings of length n, and Σ^∞ is the set of infinite binary sequences. The length of a string x is denoted by $|x|$. $<$ is the length-lexicographical ordering on Σ^* and z_n ($n \geq 0$) is the nth string under this ordering. λ is the empty string. For strings $x, y \in \Sigma^*$, xy is the concatenation of x and y. For a sequence $x \in \Sigma^* \cup \Sigma^\infty$ and an integer number $n \geq -1$, $x[0..n]$ denotes the initial segment of length $n + 1$ of x ($x[0..n] = x$ if $|x| < n + 1$) and $x[i]$ denotes the ith bit of x, i.e., $x[0..n] = x[0] \cdots x[n]$. Lower case letters $\cdots, k, l, m, n, \cdots, x, y, z$ from the middle and the end of the alphabet will denote numbers and strings, respectively. The letter b is reserved for elements of Σ, and lower case Greek letters $\xi, \eta, \cdots$ denote infinite sequences from Σ^∞.

A subset of Σ^* is called a language or simply a set. Capital letters are used to denote subsets of Σ^* and boldface capital letters are used to denote subsets of Σ^∞. The cardinality of a language A is denoted by $\|A\|$. We identify a language A with its characteristic function, i.e., $x \in A$ iff $A(x) = 1$. The characteristic sequence χ_A of a language A is the infinite sequence $\chi_A = A(z_0)A(z_1)A(z_2)\cdots$.

We freely identify a language with its characteristic sequence and the class of all languages with the set Σ^∞. For a language $A \subseteq \Sigma^*$ and a string $z_n \in \Sigma^*$, $A \upharpoonright z_n$ denotes the finite initial segment of χ_A below z_n, i.e., $A \upharpoonright z_n = \chi_A[0..n-1]$. For languages A and B, $\bar{A} = \Sigma^* - A$ is the complement of A, $A \Delta B = (A-B) \cup (B-A)$ is the symmetric difference of A and B.

We fix a standard polynomial time computable and invertible pairing function $\lambda x, y < x, y >$ on Σ^*. We will use $\mathbf{P}$, $\mathbf{E}$ and $\mathbf{E_2}$ to denote the complexity classes $DTIME(poly)$, $DTIME(2^{linear})$ and $DTIME(2^{poly})$, respectively. Finally, we fix a recursive enumeration $\{P_e : e \geq 0\}$ of $\mathbf{P}$ such that $P_e(x)$ can be computed in $O(2^{|x|+e})$ steps (uniformly in e and x).

We close this section by introducing a fragment of Lutz's effective measure theory which will be sufficient for our investigation.

Definition 1. A *martingale* is a function $F : \Sigma^* \rightarrow Q^+$ such that, for all $x \in \Sigma^*$,

$$F(x) = \frac{F(x1) + F(x0)}{2}.$$

A martingale F *succeeds* on a set $A \subseteq \Sigma^*$ if $\limsup_n F(A \upharpoonright z_n) = \infty$.

Definition 2. (Lutz [7]) A class $\mathbf{C}$ of sets has *p-measure 0* $(\mu_p(\mathbf{C}) = 0)$ if there is a polynomial time computable martingale $F : \Sigma^* \rightarrow Q^+$ which succeeds on every set in $\mathbf{C}$. The class $\mathbf{C}$ has *p-measure 1* $(\mu_p(\mathbf{C}) = 1)$ if $\mu_p(\bar{\mathbf{C}}) = 0$ for the complement $\bar{\mathbf{C}} = \{A \subseteq \Sigma^* : A \notin \mathbf{C}\}$ of $\mathbf{C}$.

Definition 3. (Schnorr [11]) A set A is n^k-*random* if, for every n^k-time computable martingale F, F does not succeed on A. A set A is *p-random* if A is n^k-random for all $k \in N$.

The following theorem is useful in the study of p-measure theory.

Theorem 4. *A class $\mathbf{C}$ of sets has p-measure 0 if and only if there exists a number $k \in N$ such that there is no n^k-random set in $\mathbf{C}$.*

Proof. See, e.g., [15]. ∎

3 Resource Bounded Randomness versus Polynomial Time Unsafe Approximations

For the reason of completeness, in this section we review the results in Wang [15, 16] which show the relations between the resource bounded randomness concept and polynomial time unsafe approximation concepts.

Definition 5. (Duris and Rolim [5] and Yesha [19]) A *polynomial time unsafe approximation* of a set A is a set $B \in \mathbf{P}$. The set $A \Delta B$ is called the *error set* of the approximation. Let f be a function defined on the natural numbers such that $\lim\sup_{n\to\infty} f(n) = \infty$. A set A is Δ-*levelable with density* f if, for any set $B \in \mathbf{P}$, there is another set $B' \in \mathbf{P}$ such that

$$\|(A\Delta B)\upharpoonright z_n\| - \|(A\Delta B')\upharpoonright z_n\| \geq f(n) \qquad (1)$$

for almost all $n \in N$. A set A is Δ-*levelable* if A is Δ-levelable with density f for some f.

Definition 6. (Ambos-Spies [1]) A polynomial time unsafe approximation B of a set A is *optimal* if, for any approximation $C \in \mathbf{P}$ of A,

$$\exists k \in N \ \forall n \in N \ (\|(A\Delta B)\upharpoonright z_n\| < \|(A\Delta C)\upharpoonright z_n\| + k) \qquad (2)$$

A set A is *weakly Δ-levelable* if, for any polynomial time unsafe approximation B of A, there is another polynomial time unsafe approximation B' of A such that

$$\forall k \in N \ \exists n \in N \ (\|(A\Delta B)\upharpoonright z_n\| > \|(A\Delta B')\upharpoonright z_n\| + k). \qquad (3)$$

It should be noted that our above definitions are a little different from the original definitions of Ambos-Spies [1], Duris and Rolim [5], and Yesha [19]. In the original definitions, they considered the errors on strings up to certain length (i.e. $\|(A\Delta B)^{\leq n}\|$) instead of errors on strings up to z_n (i.e. $\|(A\Delta B)\upharpoonright z_n\|$).

Lemma 7. *(Ambos-Spies [1])*

1. *A set A is weakly Δ-levelable if and only if A does not have an optimal polynomial time unsafe approximation.*
2. *If a set A is Δ-levelable then it is weakly Δ-levelable.*

In Wang [15, 16], we have established the following relations between the p-randomness concept and unsafe approximation concepts.

Theorem 8. *(Wang [15, 16]) The class of Δ-levelable sets has p-measure 0.*

Theorem 9. *(Wang [15, 16]) The class of sets which have optimal polynomial time unsafe approximations has p-measure 0.*

Corollary 10. *(Wang [15, 16]) The class of sets which are weakly Δ-levelable but not Δ-levelable has p-measure 1.*

Corollary 11. *(Wang [15, 16]) Every p-random set is weakly Δ-levelable but not Δ-levelable.*

4 Resource Bounded Stochasticity versus Polynomial Time Unsafe Approximations

As we have mentioned in the introduction, the first notion of randomness was proposed by von Mises [10]. He called a sequence random if every subsequence obtained by an admissible selection rule satisfies the law of large numbers. A formalization of this notion, based on formal computability was given by Church [4] in 1940. Following Kolmogorov (see [12]) we call randomness in the sense of von Mises and Church stochasticity.

For a formal definition of Church's stochasticity concept, we first formalize the notion of a selection rule.

Definition 12. A *selection function* f is a partial recursive function $f : \Sigma^* \to \Sigma$. A selection function f is *dense along* A if $f(A \upharpoonright x)$ is defined for all x and $f(A \upharpoonright x) = 1$ for infinitely many x.

By interpreting A as the infinite 0-1-sequence χ_A, a selection function f selects the subsequence $A(x_0)A(x_1)A(x_2)\cdots$ of A where $x_0 < x_1 < x_2 < \cdots$ are the strings x such that $f(A \upharpoonright x) = 1$. In particular, f selects an infinite subsequence ξ of χ_A iff f is dense along A. So Church's stochasticity concept can be defined as follows.

Definition 13. (Church [4]) A set A is *stochastic* if, for every selection function f which is dense along A and for $b \in \Sigma$,

$$\lim_n \frac{\|\{i < n \ : \ f(A \upharpoonright z_i) = 1 \ \& \ A(z_i) = b\}\|}{\|\{i < n \ : \ f(A \upharpoonright z_i) = 1\}\|} = \frac{1}{2}. \tag{4}$$

For the resource bounded version of Church stochasticity, Ambos-Spies, Mayordomo, Wang and Zheng [2] introduced the following n^k-stochasticity notion.

Definition 14. (Ambos-Spies et al. [2]) An n^k-*selection function* is a total selection function f such that $f \in DTIME(n^k)$. A set A is n^k-*stochastic* if, for every n^k-selection function f which is dense along A and for $b \in \Sigma$, (4) holds. A set A is p-*stochastic* if it is n^k-stochastic for all $k \in N$.

These concepts can also be characterized in terms of prediction functions. A prediction function f is a procedure which, given a finite initial segment of a 0-1-sequence, predicts the value of the next member of the sequence. We will show that a set A is stochastic iff, for every partial prediction function which makes infinitely many predictions along A, the numbers of the correct and incorrect predictions are asymptotically the same.

Definition 15. (Ambos-Spies et al. [2]) A *prediction function* f is a partial function $f : \Sigma^* \to \Sigma$. An n^k-*prediction function* f is a prediction function f such that $f \in DTIME(n^k)$ and $domain(f) \in DTIME(n^k)$. A prediction function f is *dense along* A if $f(A \upharpoonright x)$ is defined for infinitely many x. A *meets* *(avoids) f at x if $f(A \restriction x)$ is defined and $f(A \restriction x) = A(x)$ $(f(A \restriction x) = 1 - A(x))$. A meets f balancedly if*

$$\lim_n \frac{\|\{i < n \; : \; f(A \restriction z_i) = A(z_i)\}\|}{\|\{i < n \; : \; f(A \restriction z_i) \downarrow\}\|} = \frac{1}{2}. \tag{5}$$

Theorem 16. *(Ambos-Spies et al. [2]) For any set A, the following are equivalent.*

1. *A is n^k-stochastic (p-stochastic).*
2. *A meets balancedly every n^k-prediction (p-prediction) function which is dense along A.*

The following theorem is straightforward.

Theorem 17. *(Ambos-Spies et al. [2]) If a set A is n^k-random then it is n^k-stochastic.*

We first show that neither Δ-levelability nor optimal approximability does imply p-stochasticity.

Theorem 18. *1. There is a non-p-stochastic set B in $\mathbf{E}_2$ which has an optimal unsafe approximation.*
2. *There is a non-p-stochastic set B in $\mathbf{E}_2$ which is Δ-levelable.*

Proof. 1. Let $A \in \mathbf{E}_2$ be a set which has an optimal unsafe approximation (the existence of such A has been shown in Ambos-Spies [1]), and let $B = \{z_{2n}, z_{2n+1} : z_n \in A\}$. Then B has an optimal unsafe approximation and the prediction function f defined by

$$f(x) = \begin{cases} x[|x| - 1] & \text{if } |x| \text{ is odd} \\ \uparrow & \text{otherwise} \end{cases}$$

witnesses that B is not p-stochastic.
2. The proof is the same as that of 1. ∎

Before we prove our main theorems, we prove the following lemma which will present the basic idea underlying Ville's construction.

Lemma 19. *Let f_0, f_1 be two n^k-selection functions. Then there is a set A in $\mathbf{E}_2$ such that*

$$\|\{i < n : f_b(A \restriction z_i) = 1 = A(z_i)\}\| > \|\{i < n : f_b(A \restriction z_i) = 1 = 1 - A(z_i)\}\| \tag{6}$$

for all $n \in N$ and $b \in \Sigma$.

Proof. The construction of A is as follows.

Let $\xi_{0,0} = \xi_{0,1} = \xi_{1,0} = \xi_{1,1} = 110101010 \cdots \cdots \in \Sigma^\infty$. For $i \in N$, assume that $\chi_A[0..i-1]$ has already been defined. If $(b_0, b_1) = (f_0(\chi_A[0..i-1]), f_1(\chi_A[0..i-1]))$, then let $\chi_A[i]$ be the first bit in the sequence ξ_{b_0, b_1} that has not been used.

For the above constructed set A, every initial segment of the sequence selected by f_0 (f_1) from χ_A is a "mixture" of the initial segments of $\xi_{1,0}$ and $\xi_{1,1}$ ($\xi_{0,1}$ and $\xi_{1,1}$). Hence it satisfies the requirements of the lemma. ∎

Theorem 20. *There is a p-stochastic set $A \in \mathbf{E}_2$ satisfying the following properties.*

1. *For every p-selection function f which is dense along A, there is an unbounded nondecreasing function $r(n)$ such that*

$$\|\{i < n : f(A \restriction z_i) = 1 = A(z_i)\}\| \geq \|\{i < n : f(A \restriction z_i) = 1 = 1 - A(z_i)\}\| + r(n)$$
(7)

   for almost all $n \in N$.
2. *A has an optimal unsafe approximation.*

Proof. Let $f_0, f_1, \cdots \cdots$ be an enumeration of all p-selection functions. The construction of A is a modification of the construction in Lemma 19. The detailed construction is as follows.

Let $n_j = 2^{2j}$ for all $j \in N$, and let $\xi_w = 1110101010 \cdots \cdots \in \Sigma^\infty$ for all $w \in \Sigma^*$. For $i \in N$, assume that $\chi_A[0..i-1]$ has already been defined. Let $x = f_0(\chi_A[0..i-1])f_1(\chi_A[0..i-1]) \cdots f_{i-1}(\chi_A[0..i-1])$ and j be the least integer such that we have used less than n_j bits from $\xi_{x[0..j]}$. Then let $\chi_A[i]$ be the first bit in $\xi_{x[0..j]}$ that we have not used.

We show that the above constructed set A satisfies our requirements by establishing two Claims.

Claim 1 *Let f be a p-selection function. Then the selected subsequence by the selection function f satisfies the law of large numbers and there is an unbounded nondecreasing function $r(n)$ satisfying (7).*

Proof. The proof of the claim is exactly the same as that for the Ville's original construction, see, e.g. [12]. In the following, we will only give the outline of the intuition. The basic idea underlying the above construction is the same as that underlying the construction in Lemma 19. But here there are countably many selection rules. Whence each bit of the constructed sequence is characterized by an infinite binary sequence $b_0 b_1 \cdots \cdots$ ($b_i = 1$ if f_i selects this bit). In other words, each bit is characterized by an infinite path in a binary tree. Nevertheless, we only use an initial segment of this path. More precisely, at each stage of our construction one of the vertices of the binary tree is called *active*. To find out the active vertex we start from the root and follow the path until we find a vertex

$x_{[0..j]}$ which was active less than n_j times. Because $n_0 < n_1 < \cdots\cdots$ grows fast enough, we can ensure that the selected subsequence by the selection function f satisfies the law of large numbers (the details are omitted here, for those who have interest, it is referred to [12]). Furthermore, each base sequence is $111010\cdots\cdots$, whence it is straightforward that there is an unbounded nondecreasing function $r(n)$ satisfying (7). ∎

Claim 2 $B = \Sigma^*$ *is an optimal unsafe approximation of* A. *That is to say, for every set* $C \in \mathbf{P}$ *such that* $\|C \Delta B\| = \infty$, *(2) holds.*

Proof. Define a p-selection function f by letting

$$f(x) = \begin{cases} 1 & \text{if } C(z_{|x|}) = 0. \\ 0 & \text{otherwise.} \end{cases}$$

Then, by (7),

$$\|(A \Delta C) \upharpoonright z_n\| - \|(A \Delta B) \upharpoonright z_n\|$$

$$= \|\{i < n : f(A \upharpoonright z_i) = 1 = A(z_i)\}\| - \|\{i < n : f(A \upharpoonright z_i) = 1 = 1 - A(z_i)\}\|$$

$$> 0$$

for almost all $n \in N$. Hence (2) holds. ∎

Before proving the second main theorem of our paper, we prove a lemma at first.

Lemma 21. *Let* $B_{0,0}, B_{0,1}, B_{1,0}, B_{1,1}, B_{2,0}, B_{2,1}, \cdots\cdots$ *be a sequence of mutually disjoint sets which has a universal characteristic function in* $\mathbf{E}$ *such that* $\cup_{i \in N} \cup_{b=0,1} B_{i,b} = \Sigma^*$. *Then there is a p-stochastic set* $A \in \mathbf{E}_2$ *satisfying the following properties.*

1. *For each* $i \in N$, *let* $\alpha_{i,0} = b_0 b_1 b_2 \cdots\cdots$, *where*

$$b_j = \begin{cases} A(z_j) & \text{if } z_j \in B_{i,0} \\ \lambda & \text{if } z_j \notin B_{i,0} \end{cases}$$

   If $\alpha_{i,0} \in \Sigma^\infty$, *then there is an unbounded nondecreasing function* $r_{i,0}(n)$ *such that* $\|\{j < n : \alpha_{i,0}[j] = 0\}\| \geq \|\{j < n : \alpha_{i,0}[j] = 1\}\| + r_{i,0}(n)$ *for almost all* $n \in N$.

2. *For each* $i \in N$, *let* $\alpha_{i,1} = b_0 b_1 b_2 \cdots\cdots$, *where*

$$b_j = \begin{cases} A(z_j) & \text{if } z_j \in B_{i,1} \\ \lambda & \text{if } z_j \notin B_{i,1} \end{cases}$$

   If $\alpha_{i,1} \in \Sigma^\infty$, *then there is an unbounded nondecreasing function* $r_{i,1}(n)$ *such that* $\|\{j < n : \alpha_{i,1}[j] = 1\}\| \geq \|\{j < n : \alpha_{i,1}[j] = 0\}\| + r_{i,1}(n)$ *for almost all* $n \in N$.

Proof. Let $f_0, f_1, \cdots\cdots$ be an enumeration of all p-selection functions. The proof is a nested combination of infinitely many copies of the construction in the proof of Theorem 20. That is to say, for each $B_{i,b}$, we construct $\alpha_{i,b}$ in the same way as in the construction of A in the proof of Theorem 20. The formal construction is given below.

Let $n_j = 2^{3j}$ for all $j \in N$, and let

$$\begin{aligned}
\xi_w &= 10101010\cdots\cdots \in \Sigma^\infty \\
\xi_{w,j,1} &= 1110101010\cdots\cdots \in \Sigma^\infty \\
\xi_{w,j,0} &= 00010101010\cdots\cdots \in \Sigma^\infty
\end{aligned}$$

for all $w \in \Sigma^*$ and $j \in N$. For $i \in N$, assume that $\chi_A[0..i-1]$ has already been defined. Now we show how to define $\chi_A[i]$. Let j, b be the unique numbers such that $z_i \in B_{j,b}$. If the condition

- For all $s \leq j$ such that $f_s(\chi_A[0..i-1]) = 1$, there is a stage $u < i$ such that $f_s(\chi_A[0..u-1]) = 1$ and $\chi_A[u]$ was constructed from $\xi_{w,m,b'}$ or ξ_w for some $|w| \geq 3j$.

holds, then we construct $\chi_A[i]$ according to the following process 2, otherwise construct $\chi_A[i]$ according to the process 1.

1. Let $x = f_0(\chi_A[0..i-1])f_1(\chi_A[0..i-1])\cdots f_{i-1}(\chi_A[0..i-1])$ and s be the least integer such that we have used less than n_s bits from $\xi_{x[0..s]}$. Then let $\chi_A[i]$ be the first bit in $\xi_{x[0..s]}$ that we have not used.
2. Let $x = f_0(\chi_A[0..i-1])f_1(\chi_A[0..i-1])\cdots f_{i-1}(\chi_A[0..i-1])$ and s be the least integer such that we have used less than n_s bits from $\xi_{x[0..s],j,b}$. Then let $\chi_A[i]$ be the first bit in $\xi_{x[0..s],j,b}$ that we have not used.

In the construction, we have a base tree of binary strings where each vertex corresponds to the infinite binary sequence $1010\cdots\cdots$. And for each $B_{j,b}$ ($j \in N, b \in \Sigma$) we have a tree of binary strings where each vertex corresponds to the infinite binary sequence $111010\cdots\cdots$ if $b = 1$ and $0001010\cdots\cdots$ otherwise. At each stage of our constuction, one tree will be called *active*, and one vertex on the active tree will be called *active*. To find out the active tree, first we compute the unique numbers j, b such that $z_i \in B_{j,b}$. If the condition

- For all $s < j$ such that $f_s(\chi_A[0..i-1]) = 1$, there is a stage $u < i$ such that $f_s(\chi_A[0..u-1]) = 1$ and $\chi_A[u]$ was constructed from $\xi_{w,m,b'}$ or ξ_w for some $|w| \geq 3j$.

holds then the tree corresponds to $B_{j,b}$ will be active at stage i, otherwise the base tree will be active. To find out the active vertex on the active tree, it is the same as in the proof of Theorem 20.

For each $j \in N$ and $b \in \Sigma$, there is a number $i_{j,b}$ such that the tree corresponding to $B_{j,b}$ will be active at any stage $i > i_{j,b}$ when $z_i \in B_{j,b}$. Hence, in the same way as in the proof of Theorem 20, it is easily checked that properties 1 and 2 of the lemma are satisfied.

Now it remains to show that the above constructed set A is p-stochastic. That is to say, we need to show that each selection function f_n selects a balanced subsequence.

Let $b_0 b_1 \cdots \cdots$ be the infinite subsequence obtained by the application of the selection function f_n. Let us consider an arbitrary initial segment $b_0 b_1 \cdots b_t$ of the sequence $b_0 b_1 \cdots \cdots$ and the vertices (strings) of the binary trees corresponding to these bits. Let x be one of the longest strings among these strings (vertices) corresponding to the bits in $b_0 b_1 \cdots b_t$. Then, by the construction, the number of trees which correspond to these bits is not greater than $|x|/3$. Without loss of generality, we may assume that $|x| > n + 1$. First we give a lower bound of t as a function of $|x|$. If the string x on one tree T is used as active, then the string $x' = (x$ without the last bit$)$ on T is used as active for $2^{3(|x|-2)}$ times. The nth bit of x' is equal to 1 (we assume that $|x| > n + 1$), hence all the bits corresponding to x' is selected by f_n. So the length $t + 1$ of $b_0 b_1 \cdots b_t$ is at least $2^{3(|x|-2)}$. Now $b_0 b_1 \cdots b_t$ can be divided into two groups. For some of them the corresponding strings (vertices) have length at most n, the total number of such bits is bounded by $(2^0 2^0 + \cdots + 2^n 2^{3n}) \cdot |x|/3$, so we may ignore them. For other bits the corresponding strings (vertices) have length greater than n and the nth bit is equal to 1. So the total number of such kind of strings (vertices) used does not exceed $(1 + 2 + \cdots + 2^{|x|}) \cdot |x|/3 < 2^{2|x|}$. The difference between the number of zeros and the number of ones in each sequence corresponding to each string (vertex) is at most 3. Thus the difference between the number of ones and the number of zeros in $b_0 b_1 \cdots b_t$ does not exceed $3 \cdot 2^{2|x|}$. Hence the frequency of ones in $b_0 b_1 \cdots b_t$ is close to $1/2$ (the difference is less than $(3 \cdot 2^{2|x|})/(2^{3(|x|-2)})$ and tends to zero). ∎

Now we are ready to prove our another main theorem.

Theorem 22. *There is a p-stochastic set A in $\mathbf{E}_2$ which is Δ-levelable.*

Proof. Let $P_0, P_1, P_2 \cdots \cdots$ be an enumeration of all sets in $\mathbf{P}$. For $i \in N$ and $b \in \Sigma$, let $B_{i,b} = \{z_{<i,j>} : j \in N$ and $P_i(z_{<i,j>}) = 1 - b\}$. Let $A \in \mathbf{E}_2$ be the p-stochastic set in Lemma 21 corresponding to the sequence $B_{0,0}, B_{0,1}, B_{1,0}, B_{1,1}, B_{2,0}, B_{2,1}, \cdots \cdots$ of sets. We have to show that A is Δ-levelable. For each infinite set P_i, define a polynomial time computable set P_i' by letting

$$P_i'(z_n) = \begin{cases} 1 - P_i(z_n) & \text{if } n = <i, j> \text{ for some } j \in N \\ P_i(z_n) & \text{otherwise} \end{cases}$$

It suffices to show that (1) holds with P_i and P_i' in place of B and B' respectively. Let $\alpha_{i,0}$ and $\alpha_{i,1}$ be defined as in Lemma 21. Then at least one of them is an infinite sequence. Without loss of generality, we may assume that $\alpha_{i,0}$ is infinite and $\alpha_{i,1}$ is finite. By Lemma 21, there is an unbounded nondecreasing function $r_{i,0}(n)$ such that $\|\{j < n : \alpha_{i,0}[j] = 0\}\| \geq \|\{j < n : \alpha_{i,0}[j] = 1\}\| + r_{i,0}(n)$ for almost all $n \in N$. Hence

$$\|(A \Delta P_i) \upharpoonright z_n\| - \|(A \Delta P_i') \upharpoonright z_n\|$$
$$\geq \|\{j < n_1 : \alpha_{i,0}[j] = 0\}\| - \|\{j < n_1 : \alpha_{i,0}[j] = 1\}\| - |\alpha_{i,1}|$$
$$\geq r_{i,0}(n_1) - |\alpha_{i,1}|$$

for almost all $n \in N$, where $n_1 = \|\{j < n : j = < i, k >$ for some $k \in N$ and $P_i(z_j) = 1\}\|$. That is to say, (1) holds with P_i, P_i' and $r_{i,0}(n_1) - |\alpha_{i,1}|$ in place of B, B' and $f(n)$ respectively. ∎

Our results in this paper show that p-randomness implies weak Δ-levelability, but it implies neither Δ-levelability nor optimal approximability. However, p-stochasticity is independent of weak Δ-levelability, Δ-levelability and optimal approximability.

As a summary, we list all these relations among randomness, stochasticity and approximations. There are sets $A, B, C, D, E \in \Sigma^*$ which satisfy the properties in **Table 1**.

Table 1. The relations among randomness, stochasticity and approximations

	p-random	p-stochastic	Δ-levelable	weakly Δ-levelable	optimally approximable
A	yes	yes	no	yes	no
B	no	yes	yes	yes	no
C	no	yes	no	no	yes
D	no	no	yes	yes	no
E	no	no	no	no	yes

References

1. K. Ambos-Spies. On optimal polynomial time approximations: P-levelability vs. Δ-levelability. In *Proc. 22nd ICALP*, Lecture Notes in Comput. Sci., 944, pages 384–392. Springer Verlag, 1995.
2. K. Ambos-Spies, E. Mayordomo, Y. Wang, and X. Zheng. Resource-bounded balanced genericity, stochasticity and weak randomness. In *Proc. 13rd STACS*, Lecture Notes in Comput. Sci., 1046, pages 63-74. Springer Verlag, 1996.
3. K. Ambos-Spies, and J. Lutz. The Workshop on *Randomness and Information* in Dagstuhl, July 15-19, 1996.
4. A. Church. On the concept of a random sequence. *Bull. Amer. Math. Soc.*, 45:130–135, 1940.
5. P. Duris and J. D. P. Rolim. E-complete sets do not have optimal polynomial time approximations. In *Proc. 19th MFCS*, Lecture Notes in Comput. Sci., 841, pages 38–51. Springer Verlag, 1994.
6. K. Ko. On the notion of infinite pseudorandom sequences. *Theoret. Comput. Sci.*, 48:9–33, 1986.
7. J. H. Lutz. Almost everywhere high nonuniform complexity. *J. Comput. System Sci.*, 44:220–25, 1992.

8. P. Martin-Löf. The definition of random sequences. *Inform. and Control*, 9:602–619, 1966.

9. A. R. Meyer and M. S. Paterson. With what frequency are apparently intractable problems difficult? Technical Report TM-126, Laboratory for Computer Science, MIT, 1979.

10. R. von Mises. Grundlagen der Wahrscheinlichkeitsrechnung. *Math. Z.*, 5:52–99, 1919.

11. C. P. Schnorr. *Zufälligkeit und Wahrscheinlichkeit. Eine algorithmische Begründung der Wahrscheinlichkeitstheorie.* Lecture Notes in Math. 218. Springer Verlag, 1971.

12. V. A. Uspenskii, A. L. Semenov, and A. Kh. Shen. Can an individual sequence of zeros and ones be random? *Russian Math. Surveys*, 45:121–189, 1990.

13. J. Ville. *Ètude Critique de la Notion de Collectif.* Gauthiers-Villars, Paris, 1939.

14. A. Wald. Sur la notion de collectif dans le calcul des probabilités. *C. r. Acad. Sci. Paris*, 202:180–183, 1936.

15. Y. Wang. *Randomness and Complexity.* PhD thesis, Heidelberg, 1996.

16. Y. Wang. Genericity, randomness, and polynomial time approximations. To appear in *SIAM J. Comput.*

17. Y. Wang. The law of the iterated logarithm for p-random sequences. In Proc. 11th Conf. Computational Complexity (formerly Conf. on Structure in Complexity Theory), pages 180-189. IEEE Computer Society Press, 1996.

18. R. Wilber. Randomness and the density of hard problems. In *Proc. 24th Symp. FOCS*, pages 335–342. IEEE Computer Society Press, 1983.

19. Y. Yesha. On certain polynomial-time truth-table reducibilities of complete sets to sparse sets. *SIAM J. Comput.*, 12:411–425, 1983.

Author Index

Springer
and the
environment

At Springer we firmly believe that an international science publisher has a special obligation to the environment, and our corporate policies consistently reflect this conviction.

We also expect our business partners – paper mills, printers, packaging manufacturers, etc. – to commit themselves to using materials and production processes that do not harm the environment. The paper in this book is made from low- or no-chlorine pulp and is acid free, in conformance with international standards for paper permanency.

 Springer

Lecture Notes in Computer Science

For information about Vols. 1–1191

please contact your bookseller or Springer-Verlag